YUKON

LOS ANGELES

FLORIDA

THE CARIBBEAN

PERU

ANCASTER HIGH & VOCATIONAL SCHOOL
GEOGRAPHY DEPARTMENT

student's name	date issued	condition when issued	teacher's signature
Rachel Hunter	Sept 91	New	*(signature)*
Bob Simington	Feb 91	Next to New	LS
Taimi Clark	Sept/92	"	*(signature)*
Bonar Bulger	Feb 3/93	etc.	*(signature)*

Book Number _____ 91-20 _____

Travel & Tourism

A World Regional Geography

Travel&Tourism
A World Regional Geography

CHRISTINE HANNELL

B.Ed., M.Sc., GEOGRAPHY
Head of Geography
Centennial Collegiate and Vocational Institute
Guelph, Ontario

ROB HARSHMAN

B.Ed., M.A., GEOGRAPHY
Head of Geography
T.L. Kennedy Secondary School
Mississauga, Ontario

GRAHAM DRAPER

M.Ed., M.A., GEOGRAPHY
Head of Geography
Markville Secondary School
Markham, Ontario

John Wiley & Sons

Toronto New York Chichester Brisbane Singapore

Canadian Cataloguing in Publication Data

Hannell, Christine
  Travel and Tourism

Includes index.
ISBN 0-471-79484-8

1. Geography.   I. Harshman, Robert.   II. Draper, Graham A.   III. Title.

G128.H25 1991   910'.91   C91-094699-X

PRODUCTION CREDITS
Design: Julian Cleva
Illustration: Julian Cleva/Selwyn Simon
Typesetter: True to Type Inc.
Film: M.C.O. Graphics
Printer: Friesen Printers

Printed and bound in Canada
10 9 8 7 6 5 4 3 2 1

Dedications

To all of the people who want to learn about the world.
C.H.

To Susan
R.H.

To Sue, for her support and encouragement
G.D.

About the Cover

Open your book and lay it flat on your desk. Do you notice anything unusual about the map of the world shown there? You may feel the map is "just not right" or is "the wrong way around". That's because here, in Canada, we are used to seeing world maps that have the Atlantic Ocean as their main focus with Canada on the left and Europe on the right. Our cover map has the Pacific Ocean as its focus, with Canada on the right and Australia and Asia on the left. It is a Pacific view. Why do you think we chose to put a Pacific view map on our cover?

The cover photographs show five of the eleven regions studied in the text. Where do you think each photograph shows? Check your answers against the information given on page ix.

The eleven regions studied in this text are: Florida, Los Angeles, the Caribbean, Peru, Kenya, Egypt, Japan, the Netherlands, Thailand, Siberia, and the Yukon. Before beginning this very interesting course, prepare and complete a chart similar to the one at the bottom of the page.

Under the column headed "Region", list the eleven regions studied in the text. For the "Location" column, find the region on a map in the atlas in this text. Give two or three words and phrases that tell where the region is. The next column is headed "Thoughts Now". In this column, write down a few words or phrases that describe what you think about that region now. Then put the chart away. When you have completed the course, fill in the last column, headed "End of Course". Here write down four or five words or phrases about your thoughts on the regions now that the course is complete. Have your ideas about the regions changed? Why?

The more knowledge and understanding you have about a region, the greater your appreciation and acceptance of the people, culture, history, problems, and issues of that region will be. Write down five ways you think we can help develop and spread that type of regional understanding in people who have not been lucky enough to take this course. How can you, in your everyday life, help put your five ideas into action?

Now look at the maps on the inside covers. The one on the inside front shows the world with the eleven regions studied in this text highlighted from a Pacific centred viewpoint. The map on the inside back cover shows the same information but with an Atlantic centred viewpoint. Compare the two views before starting the course. Satisfy yourself that even though the regions look as if they are in different locations on the two maps, they only appear that way.

It is important to realize that different maps make regions appear differently in shape, size, location, and position. The regions do not change shape, size, location, or position. The different maps make them seem to do so. Look at some of the other maps provided in the atlas in this book to see how the different regions are affected by them. Understanding that maps often change or distort the information is an important part of the regional geography course.

Travel and Tourism: A World Regional Geography is designed to help you understand the world you live in a little better. We hope you enjoy the course and this text. Good luck with these studies and your future endeavours.

Region	Location	Thoughts Now	End of Course
Florida	—southeast USA —peninsula into Atlantic and Gulf of Mexico	—hot —holidays —Disney World —spring training	—retirement —investment —holidays —economic development —opportunity

Contents

To the Student

It is an understatement to say that the world is large and complex; there are times when it seems impossible to understand any of it. Often, when overwhelmed with too much information, it is helpful to focus on a part of the whole, to try to understand just a portion of it. The lessons that are learned from studying the parts can be used to gain understanding about the whole. Geographers have two approaches to focusing on parts of the whole: to look at specific topics, such as landforms, populations, or industrial activity, and to look at specific regions of the world, for example, a city, a country, or a continent. In both cases, only information that is specific to the topic or area is considered — an easier task than trying to understand the whole world at once.

This book uses the second approach to investigate the world. Eleven sample areas — better described as "regions" — are studied in some detail. Each study deals with some important characteristics of the region. While these studies will not give you a complete picture of the world, or even of the regions, much of the understanding you gain can be applied to other situations. In each study, care has been taken to consider the tourist industry and its development within that region.

The first two chapters of the book lay the foundation for your study of regions and of travel and tourism. The first chapter explains how regions are defined and ways in which this analytical tool can be used in investigations. The second chapter describes some important aspects of tourism. With this background knowledge, you will be better able to evaluate situations or circumstances found in the regions that you study in this course and beyond.

As you read *Travel and Tourism: A World Regional Geography*, think about these important questions:

- In what ways do human and physical patterns on the Earth's surface form and change over time?
- How does the study of regions and regional concepts help us to simplify the complexity of these patterns?
- What contribution can tourism make to our understanding of global patterns?
- In what ways will an improved knowledge of the world help us to be better problem solvers and decision makers?

The first two pages of each chapter consist of:

- five photographs of features found in the region;
- a map to show the region's location;
- Getting Started questions designed to draw your attention to important aspects illustrated by the photographs;
- Learning Destinations that let you know the direction that the investigations will be taking. These statements give the key themes and ideas of the chapter.

Each chapter is divided into sections. Each section is followed by Checkback and Apply questions that will help you determine if you have understood the ideas and will encourage you to use your knowledge in various ways.

Each chapter ends with Looking Back opportunities. These include:

- a review of the key ideas and vocabulary of the chapter;
- Thinking About . . . questions that allow you to check your understanding of the content;
- Atlas Activities that develop your skills in using maps;
- Further Explorations that expand your ideas.

Terms that appear in **bold type** throughout the chapters are explained in the text and are further defined in the Glossary on pages 222-224.

At various places in the chapters, you will encounter some of these features:

- The Region at a Glance: a statistical summary that will be useful when doing research;
- Case Studies: studies that focus on particular places or topics;
- Tourist Guides: information that will help travellers improve their understanding of the regions;
- Famous Tourist Destinations: descriptions of interesting places to see;
- Tourism Issues: an analysis of some important problems related to the tourist industry;
- Making Choices: activities that challenge you to apply your knowledge to make decisions.

All of these features will add to your understanding of the central themes in the chapters.

Included in the text is a 32-page atlas. The pages in this atlas are numbered A1, A2, A3… to A32. Therefore, when you are asked, for example, to turn to page A6, you would go to page 6 in the atlas section of the text.

Good luck with your investigations of world regions and travel and tourism.

Cover Photos

Netherlands

Los Angeles

Peru

Thailand

Egypt

To the Teacher

ravel and Tourism: A World Regional Geography contains, as its title indicates, two strong themes that are inextricably linked: regional studies and travel and tourism. Travel and tourism is used as a logical and interesting vehicle for studying regional geography.

The major theme of the text is investigating the world using regions as a framework. The regional approach provides a methodology for organizing information and insights about how humans use the earth's space and resources. As students gain some expertise with the processes employed in analyzing patterns and relationships, they will come to appreciate that regional generalizations provide a valuable tool for analyzing issues and problems.

Tourism, the secondary theme of this book, is a rapidly growing industry. Exploring some of tourism's potential, problems, and benefits helps students better understand important political decisions, social and economic matters, the similarities and differences among places, and regional cultural issues. As students examine the tourism issues of the regions, they become aware of how these issues shape and affect regions, their populations, and their economies. By linking tourism to regional geography, the students gain a broader ability to assess how a variety of factors come together to help shape a region.

Students are, therefore, exposed to the characteristics of the region of study, how those characteristics interrelate, are affected by, and affect other regions, and how they can be used to define a specific region. Having gained some detailed knowledge of the region, students can put what they learn about tourism and its effects into perspective. Tourism will be seen as a component of the region, one of the factors that helps to set the region apart. Within each region studied, students will be able to identify the forces that influence tourism and how tourism in turn affects the region.

Travel and Tourism: A World Regional Geography provides the basis for a course that defines and studies selected regions, gives students an opportunity for a variety of learning experiences, including planning and collaborative activities, and develops an understanding of the issues and trends facing the various regions.

Planning, decision making, problem solving, and role playing are all key elements in using *Travel and Tourism: A World Regional Geography*.

Good luck with your program.

Type of Region	Region	Page
Economic	Florida	26
City-Centred	Los Angeles	44
Cultural	The Caribbean Thailand	62 170
Historic Cultural	Peru	80
Physical	Kenya	98
River-Dependent	Egypt	116
Political	Japan	134
Demographic	The Netherlands	154
Frontier	Siberia Yukon	184 198

Acknowledgements

The authors and publishers wish to extend their thanks to Bob Goddard, R.H. Goddard Consulting, Mississauga, Ontario, who acted as the curriculum guideline consultant for this text, and to Dr. Patrick Solomon, Faculty of Education, York University, who acted as the tolerance and understanding consultant. Their contributions to this textbook are greatly appreciated.

Our thanks also to the following reviewers for their professional evaluations and thoughtful suggestions:

Frank Boddy
Innisdale Secondary School
Barrie, Ontario

Ronald Bonham
Educational Consultant
Aurora, Ontario

Don Farquharson
Geography and Environmental Education
Consultant
Durham Region Board of Education
Whitby, Ontario

Christine Zak
Vice Principal
Sir Charles Tupper Secondary School
Vancouver, British Columbia

CREDITS

Front Cover: left, Bob Haskett; right, Rob Harshman.
Back Cover: top, Andy Vammus; centre, Sheila O'Shaughnessy; bottom, Bob Haskett.

Chapter 1: Figs. 1.a and 1.b, Geographical Visual Aids; Fig. 1.c, UNICEF/Bernard Wolff; Fig. 1.d, Geographical Visual Aids; Fig. 1.e, UNICEF/Alan Harbour; Fig. 1.2, Netherlands Board of Tourism; Fig. 1.5, Barbados Board of Tourism; Fig. 1.7, Japan National Tourist Organization.

Chapter 2: Fig. 2.a, The Egyptian Tourist Authority; Fig. 2.b, Rob Harshman; Fig. 2.c, Netherlands Board of Tourism; Fig. 2.d, Yukon Government Photo; Fig. 2.e, Jane Goddard; Fig. 2.9, Consulate General of Japan; Fig. 2.10, Antigua Department of Tourism and Trade; p. 23, © 1989 by Sidney Harris-Condé Nast Traveler Magazine.

Chapter 3: Fig. 3.a, Florida Department of Commerce, Division of Tourism; Figs. 3.b, 3.c, and 3.d, Rob Harshman; Fig. 3.e, Florida Department of Commerce, Division of Tourism; Fig. 3.1, Rob Harshman; Fig. 3.4, Florida Department of Commerce, Division of Tourism; p. 34, left, Canapress Photo Service; Figs. 3.10, 3.11, and 3.14, Florida Department of Commerce, Division of Tourism.

Chapter 4: Figs. 4.a, 4.b, 4.c, 4.d, 4.e, 4.1, Rob Harshman; Fig. 4.4, US Geological Survey; Fig. 4.7, Robert Landau/First Light; Figs. 4.9, 4.11, and 4.14, Rob Harshman.

Chapter 5: Fig. 5.a, Rob Harshman; Fig. 5.b, Bob Haskett; Fig. 5.c, Jamaican Tourist Board; Fig. 5.d, Jamaican Tourist Board; Fig. 5.e, Bob Hartnell; Fig. 5.4, Jamaican Tourist Board; Fig. 5.8, Rob Harshman; Fig. 5.9, Lance Goddard; Figs. 5.13 and 5.14, Jamaican Tourist Board.

Chapter 6: Figs. 6.a, 6.b, 6.c, 6.d, and 6.e, Rob Harshman; Fig. 6.1, Sheila O'Shaughnessy; Fig. 6.2, Rob Harshman; Fig. 6.3, Consulate General of Peru; Fig. 6.5, top, Sheila O'Shaughnessy, bottom, Consulate General of Peru; Fig. 6.6, Rob Harshman; Figs. 6.7, 6.8, and 6.9, Sheila O'Shaughnessy.

Chapter 7: Figs. 7.a, 7.b, 7.c, 7.d, and 7.e, Bob Haskett; Figs. 7.2 and 7.7, Bob Haskett; Fig. 7.8, Rob Harshman; Fig. 7.16, Gerard Champlong/The Image Bank Canada.

Chapter 8: Figs. 8.a, 8.b, 8.c, 8.d, 8.e, 8.1, 8.4, 8.7, 8.8, 8.9(a), 8.9(b), 8.10(a), 8.10(b), and 8.10(c), Bob Haskett.

Chapter 9: Fig. 9.a, C. Batten; Figs. 9.b, 9.c, and 9.d, Rob Harshman; Figs. 9.e, 9.1, 9.2, 9.3, 9.4, and 9.7, Consulate General of Japan; Fig. 9.10, NKK Corporation; Fig. 9.13, Japan Railways Group; Figs. 9.15 and 9.16, C. Batten.

Chapter 10: Figs. 10.a and 10.b, Bob Haskett; Fig. 10.c, Netherlands Board of Tourism; Figs. 10.d and 10.e, E. te Bokkel; Fig. 10.1, Rob Harshman; Figs. 10.4, 10.5, and 10.6, Netherlands Board of Tourism; Fig. 10.8, Peter Haskett; Figs. 10.9 and 10.11, Netherlands Board of Tourism.

Chapter 11: Figs. 11.a and 11.b, Bob Haskett; Fig. 11.c, Chris Hannell; Fig. 11.d, Bob Haskett; Fig. 11.e, Chris Hannell; Fig. 11.1, Lily Marcinek; Figs. 11.2 and 11.3, Rob Harshman; Figs. 11.5 and 11.6, Bob Haskett; Figs. 11.7 and 11.8, Rob Harshman; Fig. 11.9, Bob Haskett.

Chapter 12: Figs. 12.a, 12.b, 12.c, 12.d, 12.e, 12.1, 12.5, 12.6, 12.7, 12.8, and 12.9, Tass/Sovfoto.

Chapter 13: Figs. 13.a, 13.b, 13.c, 13.d, and 13.e, Yukon Government Photo; page 200, excerpt from *Klondike: The Last Great Gold Rush, 1896-1899*, courtesy of Pierre Berton; Fig. 13.1, Yukon Archives; Fig. 13.3, Yukon Government Photo; Fig. 13.6, Richard Hartmier; Figs. 13.9 and 13.10, Yukon Government Photo; Fig. 13.11, Chris Hannell.

Chapter 14: Figs. 14.a, 14.b, 14.c, 14.d, and 14.e, Geographical Visual Aids.

1

Regions

AN INTRODUCTION

Figure 1.a   *Amsterdam Harbour*

Figure 1.b   *Petroleum Refinery, Curacao*

Figure 1.c   *A Market in Cairo*

Figure 1.d   Laguna Hills, Los Angeles

Figure 1.e   The Mountains of Peru

GETTING STARTED

1. Write two statements for each photograph that describe the physical environment shown. Include references to the landforms, vegetation, and climate of the places.
2. List all the economic activities that are shown or suggested by the photographs. Which of these activities are related to tourism? How?
3. Decide on two categories into which you could group the scenes shown by the photographs. Describe the criteria you used to establish your categories and list the photographs you would include in each.
4. What characteristics make each place shown in the photographs different from the others?

LEARNING DESTINATIONS

By the end of this chapter, you should be able to:
- describe how regions are defined;
- identify various types of regions;
- recognize the value of regions in understanding global patterns.

Introduction

Knowledge of places, whether close to home or far away, can be at different levels. You might know, for example, just a simple fact about a place — the Eiffel Tower is in Paris. On the other hand, you might know a good deal about Paris — that you can see the Seine River from the Eiffel Tower, that you walk up the Champs-Elysées towards the Arc de Triomphe, that the shops of Rue St. Honoré are interesting places to browse, that Paris is the cultural and economic heart of France and the hub of a national railway network. In this second case, you would have a more complex image of Paris in your mind, and your geographic knowledge would be richer in details and connections.

It is useful to think about geographic knowledge at different levels of complexity. The simplest level will be isolated facts about **locations**. For instance, you may know the location of, and be able to name, the street corner on which a large, important store in your community is located.

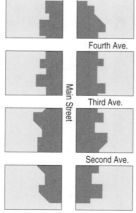

Figure 1.1a *Location*

When you see **patterns** in locations, your geographic knowledge is more detailed. You might have recognized that along the length of the major street on which the store is located, there are a series of busy and important street corners.

Figure 1.1b *Patterns*

Closer inspection of the major street may reveal some **connections** along its length. The street may serve as a bus route; shops and businesses may line the street, with larger commercial buildings occupying the locations at the street corners.

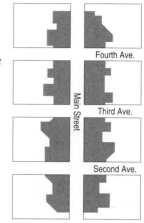

■ Commercial Activity

Figure 1.1c *Connections*

Looking beyond the major street, you may find **relationships** that this area has with the surrounding places. For example, just a few doors off the main street, there may be homes and schools, instead of businesses; that is, the land use is quite different but related, in that employers, employees, and customers would live nearby.

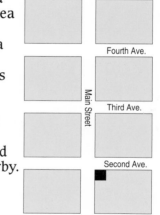

■ Commercial Activity
☐ Residential Uses

Figure 1.1d *Relationships*

By combining information that you observe about an area's patterns, connections, and relationships, you identify a **region**. This is a commercial region of shops and businesses, quite distinct from the residential region nearby.

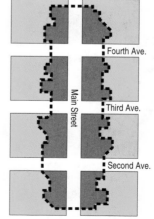

■ Commercial Activity
☐ Residential Uses
- - - Boundary of Commercial Region

Figure 1.1e *A Commercial Region*

Figure 1.2 *A Commercial Area in Amsterdam*
How might defining this area as a region help us to understand it?

Regions are a complex form of geographic knowledge. In order to understand a region, we must know what is unique about it; that is, we must know what criteria have been used to define it and establish its boundaries. The act of defining it as a region means that we set it apart from the rest of the world. Once separate, we can examine a region in greater detail, without being distracted by irrelevant facts and ideas about neighbouring regions.

Regions can help us to simplify the world and make it easier to understand. In the example we have just examined, we have identified a region of commercial activity. Likely, we will start referring to this region by a name, such as the Main Street Commercial Region or the Central Business District, as though all parts of it have similar characteristics. By doing this, we have made a generalization. We might now compare or contrast this commercial region with other commercial areas, either in the same community or elsewhere. The concepts we formed by observing this one region can be applied to other situations, helping us to simplify a very complex world.

CHECKBACK AND APPLY

1. Think about your own neighbourhood and write a statement about its:

   a) location;

   b) pattern, including boundaries;

   c) connections with other areas or regions;

   d) relationships to other areas or regions;

   e) characteristics as a region.

   Draw a simple sketch map to accompany your answer.

2. In what way is understanding a region more difficult than understanding a location, pattern, connection, or relationship?

3. We use generalizations to simplify the world in a number of different ways. Use generalizations to describe the cars that are parked in the school parking lot on a typical day.

Types of Regions

A region is a device people use to understand the world more easily. By breaking the surface up into a number of smaller units, we make the study of the world simpler. Regions fall into two types: those that have similar characteristics throughout, and those that are centred on a place or thing.

In the example discussed previously, the region has a common characteristic — commercial activities. The line marking the edge of the region (the boundary) is drawn where shops and businesses end and housing begins. This is an example of a **homogeneous region** (sometimes called a formal region), where the characteristics are similar throughout. Parks and recreation areas might also be considered to be homogeneous regions; the land is reserved for leisure use, much different from other activities that might be going on around the community.

The second type of region is called a **functional region**. Functional regions have a centre point around which some activity or functions take place. A school is a good example of a functional region. Students travel from their homes to the school; the line on a map that contains all the students' homes marks the boundary of the region. Figure 1.3 lists some functional and homogeneous regions that are found in most communities.

Figure 1.3 *Types of Regions*

Homogeneous Regions	Functional Regions
a commercial area	a school attendance area
a housing subdivision	a pizza delivery area
a park	a police district
an industrial area	a newspaper delivery route

Regions can be defined by one or more characteristics. Each of the regions identified in Figure 1.3 is defined by just one criterion — the presence of commercial activities, the homes of students who go to a particular school, and so on. There are times, however, when it is important to consider more than one criterion. Suppose, for instance, that merchants in the commercial area in our example wanted to argue for more police protection in certain areas. They might define a homogeneous region that has:

a) commercial activity;

b) high crime rates; and

c) inadequate street lighting.

The area that has all three characteristics will be smaller than the region based on commercial activity alone.

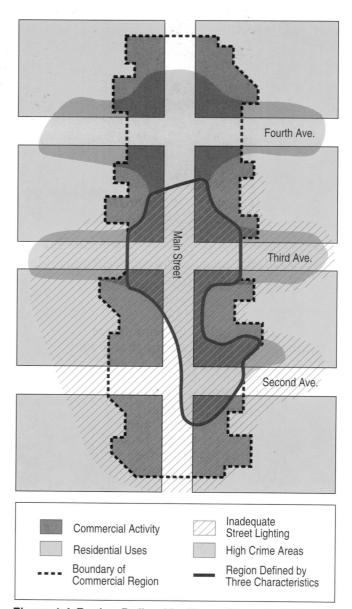

■ Commercial Activity		⫽ Inadequate Street Lighting	
Residential Uses		High Crime Areas	
---- Boundary of Commercial Region		— Region Defined by Three Characteristics	

Figure 1.4 *Region Defined by Three Characteristics*
A region defined by several characteristics is usually smaller than a region defined by just one.

4. a) Describe a homogeneous region that is found in your area. Estimate the size of the region.

b) Describe a functional region in your area. What is the centre of the region?

5. a) Identify three homogeneous regions that are found elsewhere in the world. At least one of the regions should be defined by more than one characteristic.

b) Identify three functional regions in the world. At least one should be defined by two or more characteristics.

6. What would you consider to be the greatest advantage in using regions to understand the world? Give examples to support your opinion.

Regions of Different Scales

The examples of regions that have been used up to now have all been of areas in cities. But, we can define regions of other sizes or at other scales.

Small-scale regions are those that contain small areas, such as parts of communities or just portions of neighbourhoods. In a residential subdivision (which may be defined as a region when dealing with the whole community), you might observe several types of housing. The townhouses could be defined as a region, quite separate from the single, detached homes. Similarly, a small forest might be divided into regions based on the types of trees in different parts of the forest. At this scale, the differences in regions could be quite small.

Figure 1.5 *A Farming Area in Barbados*
How might this area be considered as a large-scale region? a small-scale region?

Large-scale regions may include huge areas and have many differences in characteristics. For instance, the Middle East could be defined as a region because the major religion in the area is Islam; however, there are important political, cultural, and economic differences in this part of the world. There are times when it is appropriate to use regions of this magnitude, when the differences within the area are less important than the similarities.

The scale that is used to define regions is determined by the purpose for using them. When trying to make sense of the whole world, large-scale regions may yield more realistic understandings. Dealing with patterns and relationships within communities calls for the use of small-scale regions. That is, regions are tools to be used by inquiring minds. They do not exist on their own, but are created for the purpose of organizing information and ideas about places.

The case studies in this book are at a variety of scales, ranging from a city region to an international region:

Scale	Regions
City	Los Angeles
Sub-national	Florida Siberia Yukon
National	Peru Kenya Egypt Japan Netherlands Thailand
International	Caribbean

The choice of scale for each region was based on the objectives set for the chapter.

CHECKBACK AND APPLY

7. Your home is located within a number of different regions, such as the reception area for television signals, the sales territories for different products, and a climatic region. Make a list of eight regions of different scales within which your home will fall. Identify each as a homogeneous or functional region. You may find the atlas section of this book useful for this task.

8. Describe the regional scale that will be most appropriate for each of the following tasks. Give reasons for your choices.

   a) understanding the causes of military conflicts in the world

   b) investigating the increase in violent crimes in a community

   c) planning the most efficient way to distribute election campaign materials within a riding

   d) assessing the tourist potential of tropical places

   e) setting up an airline schedule between North America and Europe

Defining Regions

Before the boundaries of a region can be established, the criteria used to define the region must be clear. In the example already used in this chapter, the region was defined by the presence of commercial activities. All locations that had offices or shops were part of the region, all places with other land uses were not considered part of the region — they lay outside the region's boundaries. Defining this region was straightforward. But, what if the task is not so simple? Suppose you wanted to define a region by population density. Is 100 persons/km^2 a high density, or is 1000 persons/km^2? The answer depends on the reason why you are creating the region in the first place. The actual criteria used are less important than is the fact that you clearly establish your criteria to begin with and use those criteria in defining your region and establishing its boundaries.

Scale becomes a consideration when defining regions. At the community level, a region based on commercial activity is practical, but this criterion would not be useful at the national or international scale. Other criteria have to be applied to investigations at larger scales.

A wide variety of criteria can be used to define regions. The choice is dependent upon the theme or topic of investigation. In a study with an economic focus, the following criteria might be used:

- levels of income
- types of natural resources
- transportation facilities
- economic activities
- imports and exports

Other criteria would be used to examine the physical, cultural, political, and demographic character of places. Figure 1.6 shows both the focus and the characteristics that have been used to define the regions considered in this book.

Figure 1.6 *Regions Examined in This Book*

Region	Focus	Characteristics
Florida	An Economic Region	the production of wealth is well organized, and focused on tourism
Los Angeles	A City-Centred Region	the city dominates activities over a wide area
The Caribbean	A Cultural Region	the history, beliefs, and traditions of the people make this area unique
Peru	A Historic Cultural Region	the dominant cultural characteristics have developed over thousands of years
Kenya	A Physical Region	the physical environment has a significant influence on the lives of the inhabitants
Egypt	A River-Dependent Region	life is centred on the river and its valley
Japan	A Political Region	people who see themselves as separate from others and the country has developed on that attitude
The Netherlands	A Demographic Region	a densely populated area, where the people have responded to this fact in interesting ways
Thailand	A Cultural Region	the culture of this area has flourished for many hundreds of years
Siberia	A Frontier Region	a sparsely populated area, where people are settling the land and exploring its natural resources
Yukon	A Frontier Region	an area where settlement and development have occurred relatively recently

Defining a region is just the first step in investigating a place: it puts boundaries around the area to be studied and establishes the focus for investigation. Data gathering can then be organized and directed towards identifying the important locations, patterns, connections, and relationships, and producing a coherent description of the region. Our understanding of the region is based on how the locations, patterns, connections, and relationships work together to give the region its unique or special characteristics.

CHECKBACK AND APPLY

9. Explain why it is important to clearly define a region.

10. **a)** If you were to do a regional study of your school district, which focus would provide the best understanding of its characteristics: physical, economic, cultural, demographic, or something else? Give reasons for your choice.

   b) List aspects of your school district that you would examine in a regional study.

11. Describe the criteria that you might use to define regions that will enable you to investigate the following problems:

   a) poverty in a large city

   b) which foreign countries should get assistance

   c) where flood control dams should be built

   d) a good location to build a large amusement park

   e) the best place to vacation during the winter

The World
AT A GLANCE

Areas of Continents

North America	24 350 000 km²
South America	17 870 000 km²
Europe	9 840 000 km²
Asia	45 100 000 km²
Africa	30 300 000 km²
Australasia	8 550 000 km²
Antarctica	13 990 000 km²

Percentage of World Population

North America	8.1
South America	5.5
Europe	13.2
Asia	60.3
Africa	12.4
Australasia	0.5
Antarctica	- -

Per Capita GDP* (1987)

North America	US$18 110
South America	US$ 1 850
Europe	US$ 9 700
Asia	US$ 1 170
Africa	US$ 610
Australasia	US$ 8 440
Antarctica	- -

*Gross Domestic Product

Using Regions to Investigate Issues

The goal in studying regions is to obtain a coherent image of those areas. But, merely describing the region is not enough. The knowledge gained should be used to investigate problems and issues within that region, and to understand similar issues in other places. Knowledge is most useful when it is applied in ways that help solve problems.

Problems and issues often arise when change occurs in a region. Population growth may cause some recreational facilities to be overcrowded, or economic development may cause conflicts over the best use of resources. People who study the connections and relationships within a region will have insights into the nature of the problems and issues.

The insights gained from investigating regions can be applied to planning for solutions, or to avoid future problems. Planning involves making informed decisions about how space and resources are to be used. Those who have studied regions will recognize the problems that require immediate or long-term action, and the implications of not responding to situations correctly. The knowledge and understanding that come through investigation of regions, and the application of that knowledge and understanding, make the study of regions so important.

CHECKBACK AND APPLY

12. What are the advantages of using regional techniques to investigate problems? What might be some disadvantages?

13. Identify three problems or issues in your local area that might be better understood through doing a regional study. What information would you require to gain insights into each problem?

Figure 1.7 *A Scene in Tokyo*
What urban problem does this photograph show?

Chapter Review

- A region is an area of the Earth's surface that has one or more characteristic in common.
- A region is the sum of all the locations, patterns, connections, and relationships that take place within it.
- Regions may be homogeneous or functional, and may be at a wide variety of scales.
- Regions may be defined using a number of different criteria, including political, economic, cultural, physical, and demographic characteristics.
- To be useful, knowledge of regions should be applied to solve problems and to plan for the future.

Vocabulary Review

connection
functional region
homogeneous region
location
pattern
planning
region
relationship

Thinking About . . .

1. **a)** In your own words, compare homogeneous and functional regions.

   b) Compare large-scale and small-scale regions using a chart similar to the one shown in Figure 1.8.

Figure 1.8 *Types and Scales of Regions*

	Large-Scale Regions	Small-Scale Regions
Homogeneous Regions		
Functional Regions		

For each quadrant, give three examples of questions that might be investigated using regions of those types.

2. Use a flow diagram or illustration to show the steps that must be taken to study a region. The last step in the process will be planning to solve or prevent problems.

3. Consider your own province. In what ways can it be considered a homogeneous region? In what ways is it a functional region?

4. Identify some ways that tourists might use regional techniques to:

   a) identify possible destinations;

   b) plan suitable clothing and equipment for their visits;

   c) learn something of the history and culture of their destinations;

   d) identify problems or issues that might be facing tourist destinations.

5. What are some factors that will affect the way regions are defined? Describe the effect each of the following factors might have on the way researchers choose to define their regions:
   - political boundaries
   - personal knowledge
   - physical features, such as water bodies
   - available information

LOOKING BACK AT REGIONS

 Atlas Activities

1. Refer to page A1 (page 1 in the atlas section of this book).

   a) What is a hemisphere? In what ways can a hemisphere be considered a region?

   b) The amount of land varies from one hemisphere to another. Compare the northern and southern hemispheres. Which hemisphere has the greater proportion of land? Which has the greater proportion of water?

   c) What similarities do you note between the eastern and western hemispheres? In your answer, refer specifically to the patterns of land and water.

2. **a)** Turn to page A10. Identify three regions of the world that would fit into each of the following categories of quality of life: High, Medium, and Low. What criteria are used to define High, Medium, and Low qualities of life?

   b) Make several general statements about the patterns of quality of life in the world.

   c) In your own words, explain relationships among quality of life and education, literacy, language, and agriculture.

3. Use pages A10-A11, A26-A27, and A30-A31 to compare North and South America. Construct a comparison chart to show climate, vegetation, population distributions, resources, and quality of life. Write a paragraph giving a conclusion or generalization about the continents.

Further Explorations

1. **a)** Research a local region. Choose one of the regions from the following list, or another approved by your teacher.

   - the circulation area of a local newspaper
   - the area within which you can make local telephone calls
   - the delivery area for a fast-food franchise
   - your school attendance area
   - a climatic region
   - a forested area, marsh, or other distinct land area
   - an urban area
   - an industrial area

   b) Map the boundaries of your region.

   c) Write a short account that explains:

   - why this area can be considered a region;
   - why this is a functional or homogeneous region;
   - the criteria that you used to establish the boundaries of the region;
   - one issue or problem that could be examined using the criteria you established.

2. Compile ten examples of ways in which regions have been used. Cut maps and articles out of magazines and newspapers. Organize the examples into categories, such as those with an economic focus, political focus, and so on. For each article, note the scale of the region and if the region is defined based on only one characteristic, or if several have been used. On the basis of your evidence, write a conclusion about the way regions are used in everyday life.

3. Using the climatic region map on page A12, choose the largest single climatic zone and decide how you could divide it into two climatic zones. Where would the new boundary be located? What criteria would you use to decide? Map your new climatic regions, and explain why you mapped the new boundary where you did.

2

Travel and Tourism

AN INTRODUCTION

Figure 2.a   *The Pyramids, Egypt*

Figure 2.b   *The Epcot Center Florida, U.S.A.*

Figure 2.c   *Bollenveld, The Netherlands*

Figure 2.d   *White Pass, The Yukon, Canada*

GETTING STARTED

1. Examine the photos carefully. List two criteria that could be used to define each of the regions shown in the photos.
2. List two tourism-related activities that could take place in each region shown. How will these activities benefit the places in which they occur?
3. Give five reasons why people travel as tourists.
4. Which of the tourist destinations shown in the photos would you most prefer to visit? Give reasons for your answer.
5. On a global scale, what are the benefits of tourism? What problems might tourism create or make worse?

LEARNING DESTINATIONS

By the end of this chapter, you should be able to:

- recognize types of travellers;
- understand why tourism has grown so rapidly in the twentieth century;
- identify the areas of the world that benefit most from tourism;
- appreciate the effects of tourism on the economies of countries;
- recognize the importance of tourism, both economically and culturally, to the people of the world.

Introduction

T ourism on the scale that it is practised today is a relatively recent phenomenon. Until the second half of the twentieth century, tourist travel was limited to the wealthy or the adventurous. Most people seldom travelled for pleasure more than a few hours from their homes. Today, tourism is a US$2.3 trillion business and the fastest growing industry in the world. Over 125 nations rate tourism as a major industry. The reasons for this growth, and its effects, are fascinating to study.

Not all travel is tourist travel. Figure 2.1 shows that people travel for many reasons. A tourist is commonly described as someone who travels from home for more than 24 hours, but for less than one year. This includes not only people who travel for recreation, but also people who travel on business, for religious reasons, and to study. Domestic tourists stay within their own country; international tourists cross national boundaries and travel in foreign countries. Figure 2.2 gives a good indication of the rate of growth in international tourism for the period 1960-1988.

Figure 2.1 *Classification of Travellers*
Not included in the definition of "tourist" are people who change their residence and those who travel for less than a day.

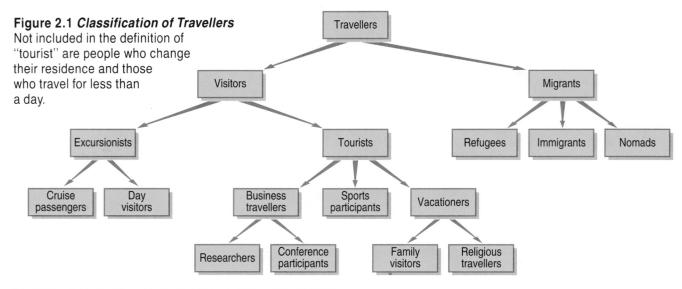

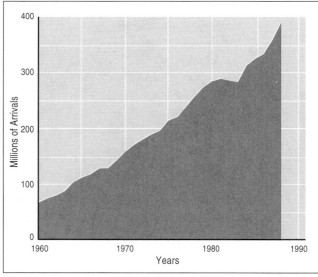

Source: World Tourism Organization

Figure 2.2 *International Tourist Arrivals, 1960-1988*
What might account for the downturn in tourism in the early 1980s?

CHECKBACK AND APPLY

1. Define the following terms:
   - traveller
   - excursionist
   - tourist
   - migrant

2. Suggest four factors that have contributed to the growth in international tourism. Explain how each factor has done so.

3. Some people have suggested that international tourism could lead to greater world peace. Do you agree or disagree with this statement? Explain your reasoning.

World Patterns of Tourism

While international tourism has grown quickly, it has not done so equally around the world. A high proportion of tourists are from European and North American countries. These areas are also the destinations of the majority of tourists, as Figures 2.3 and 2.4 indicate.

Figure 2.3 *International Tourist Arrivals by Area, 1988*

Area	Number of Tourist Arrivals (thousands)	Percentage of World Total	Percentage of Change Over 1987
Africa	12 000	3.1	18.7
Americas	72 550	18.6	7.0
East Asia and Pacific	42 000	10.8	17.8
Europe	251 500	64.5	7.5
Middle East	9 000	2.3	8.2
South Asia	2 950	0.8	2.0

Source: World Tourism Organization

Figure 2.4 *International Tourist Flows and Spending*
What factors influence the amount of tourist spending in a country?

Several factors help to explain why these areas of the world are the biggest spenders on tourism.

- **Discretionary income** has increased greatly in some countries. Discretionary income is the amount of money left over after taxes have been paid and the essentials of life (food, clothing, and shelter) purchased.

- The length of the work week has decreased and, consequently, the amount of leisure time has increased. In addition, government legislation in most European and North American countries ensures that workers get a minimum of two weeks of vacation time per year.

- Transportation methods have improved, allowing tourists to travel more conveniently at lower costs to more places.

- The travel industry has developed new vacation packages that satisfy the needs and wants of a wide range of tourists.

- Levels of education have increased, and people are more aware of travel opportunities and the value of travel experiences.

- Generally, the value placed on travel and experiencing new places has increased.

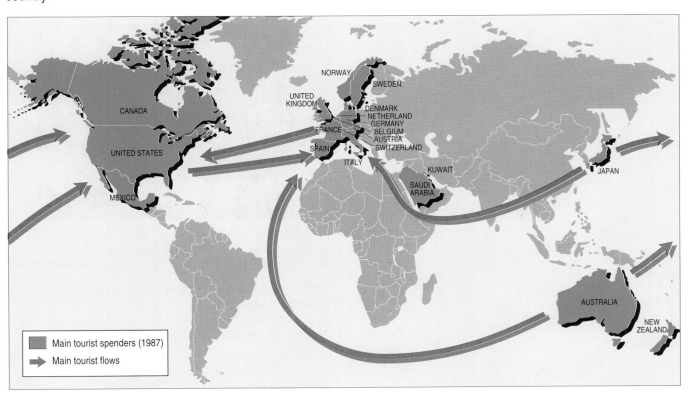

NORWAY
SWEDEN
UNITED KINGDOM
DENMARK
NETHERLAND
GERMANY
FRANCE
BELGIUM
AUSTRIA
SPAIN
SWITZERLAND
ITALY
KUWAIT
SAUDI ARABIA
JAPAN
CANADA
UNITED STATES
MEXICO
AUSTRALIA
NEW ZEALAND

Main tourist spenders (1987)
Main tourist flows

Location plays an important part in determining visitors' choices of destinations. Costs of travel, travel time, and **accessibility** (the ease with which destinations can be reached) will encourage tourists to vacation in places that are relatively close to home. For example, about two-thirds of tourists from the United States vacation in Canada and Mexico. Eighty-six percent of all Canadian visits to other countries are to the United States. In Europe, cross-border trips by car account for 20 percent of world tourism.

Worldwide, travel to neighbouring countries accounts for about three-quarters of all tourist travel.

Figure 2.5 *United States International Tourism, 1987*

Departures	Destination/Origin	Arrivals
13 245 000	Canada	12 410 000
13 500 000	Mexico	5 935 000
13 800 000	Overseas	10 505 000
6 200 000	• Europe	4 705 000
780 000	• South America	940 000
740 000	• Central America	330 000
3 850 000	• Caribbean	920 000
1 960 000	• Asia/Middle East	2 765 000
40 000	• Africa	125 000
500 000	• Australasia	415 000

Source: World Tourism Organization

4. a) Suggest five factors that will influence tourists' choice of a tourist destination.

b) Of the five factors you suggested in a), which ones can be changed or controlled by the tourist industry? Explain.

5. What areas of the world have relatively small rates of tourism? Suggest reasons to explain why this is the case.

6. a) Examine Figure 2.6. During what period of time did the greatest change in the length of the work week of Americans take place?

b) What reasons can you offer to explain the decrease in the length of the average American work week?

c) Suggest some effects that a shorter work week will have on:

• leisure time activities;
• lifestyles;
• vacation travel.

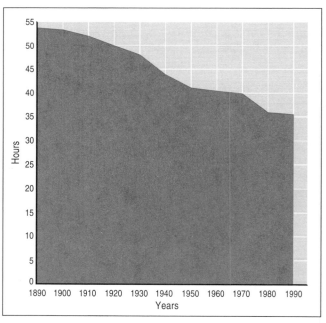

Source: U.S. Bureau of the Census

Figure 2.6 *Length of Work Week for Average American Workers, 1890-1990*
What might be some of the causes and consequences of this change in the length of the work week?

The Effects of Tourism

The tourist industry has many components. It is not a single industry, but rather a complex of interrelated businesses that serve people who are travelling. Some businesses are directly involved with the travelling public, such as airlines, hotels, theme parks, and so on. Other businesses are less directly involved, for example, financial institutions that lend money for tourist developments, government offices that regulate the industry, and manufacturers of leisure products. Some of the money spent by tourists stays in the country of origin, some is spent in transit, and some is spent in the destination country.

The money that tourists spend in their destinations has significant effects on the economies of those destinations. Although a tourist may spend money on only a small number of items, that money is spent and respent throughout the economy. For example, when a visitor buys a souvenir from a small shop, the shopkeeper, in turn, spends that money on a number of items, such as rent, salaries, supplies, and food. The money spent on rent is used by the property owner in many ways, and so it continues. This process of money being spent around the economy, creating jobs as it goes, is called the **ripple effect** (or multiplier effect).

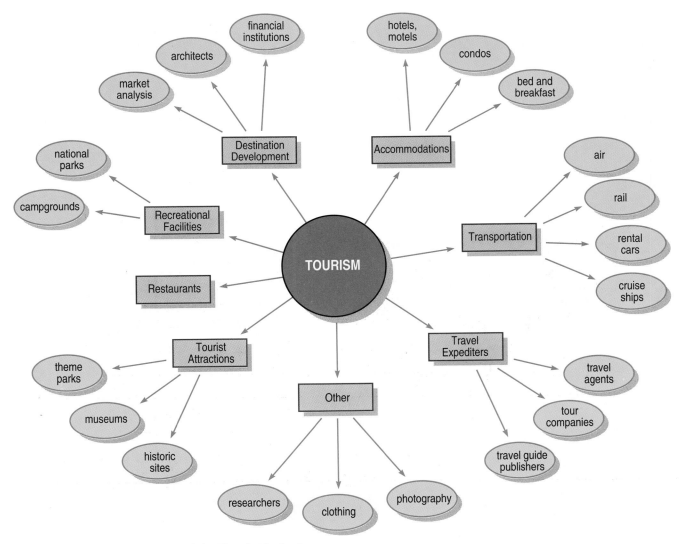

Figure 2.7 *The Components of the Tourist Industry*
Which part of the industry will likely earn the greatest share of the tourist dollar?

In addition to job creation, the tourist destination may also benefit in other ways. Governments often earn revenues through special tourist taxes and fees, which may pay for services such as schools and hospitals; large tourist-oriented facilities may make it possible for communities to construct better roads or water supply systems; governments of developing countries rely on tourism to offset the costs of imports in order to improve their balance of trade situations; tourist jobs help to diversify the economies of most countries, especially those dependent on primary industries.

Critics of tourism, however, argue that the costs far outweigh the benefits. They point to a variety of problems, not the least of which is crowding of local sites, such as beaches and historic buildings. Much of the income earned by tourism does not reach the local people.

Figure 2.8 *How Tourist Spending Spreads Throughout the Economy*

What a Tourist Spends Money On	What the Travel Industry Spends Money On	Others in the Economy Who Benefit
accommodation	wages	hotel staff
meals and beverages	taxes	doctors
transportation within the country	office/administration	gardeners
	entertainment	plumbers
sightseeing	food, beverages, and and supplies	pharmacists
entertainment		store personnel
souvenirs/gifts	legal services	farmers
clothing	electricity, gas, and water	truckers
tips	insurance	taxi drivers
medication	repairs and improvements to existing properties	civil servants (in the areas of health, roads, housing, etc.)
personal care items	interest and principal on borrowed money	painters
		carpenters
	advertising	movie theatre personnel
	return to investors	real estate agents

Figure 2.9 *A Rice-Planting Festival in Japan*
In what ways do tourist destinations benefit from tourism? In what ways is tourism in these places a problem?

Often, it is spent on imported goods, such as food and equipment, and returned to the foreign owners of the tourist facilities as profit. Expensive facilities developed for tourists — airports, for example — may not be much used by the local people. The same may be true of boutiques, restaurants, and casinos.

The greatest concern for opponents of international tourism is its effect on local cultures. The tourists' wealth is often in direct contrast to the local peoples' way of life, and may create envy and frustration. Traditional arts and crafts may be set aside as local people pursue new activities more in demand by tourists. Festivals and traditional forms of celebration can become commercialized for tourists and, as a consequence, become less meaningful to the local people. Overall, critics are concerned that in the interchange between tourists and locals, the culture of the locals suffers the greatest effects.

CHECKBACK AND APPLY

7. a) How could tourism be related to each of the following jobs?

- a performing artist
- a police officer
- a bank teller
- a salesperson in a clothing store

b) List five jobs in your area that are directly related to tourism, and five that are indirectly related.

8. a) Explain this statement in your own words: "What is built for tourists benefits everyone."

b) Do you agree with the statement? Give reasons to support your opinion.

9. A number of countries in the world have encouraged tourism in spite of the risks to the local cultures. In these cases, the economic benefits were considered to be more important than any negative cultural effects.

a) What negative effects might the development of the tourist industry have for these cultures?

b) Suggest some ways in which the interaction between the local cultures and the cultures of the visitors might benefit the people of these countries.

c) Suggest three strategies that could minimize the harmful effects of outside cultures.

Tourism

AT A GLANCE

Cost of a Weekend Stay in Selected Cities

	US$
Amsterdam	$373
Cairo	$187
Hong Kong	$284
Los Angeles	$238
Mexico City	$202
Nairobi	$206
Rio de Janeiro	$228
Tokyo	$631
Toronto	$296

Tourist Receipts for Top Ten Countries, 1987 (US$ millions)

United States	$15 374
Spain	$14 760
Italy	$12 174
France	$12 008
United Kingdom	$10 229
West Germany	$ 7 716
Austria	$ 7 604
Switzerland	$ 5 352
Canada	$ 3 939
Mexico	$ 3 497

World-Wide Capacity of Hotels, 1986

Europe	50.2%
North America	28.8%
Asia	10.4%
Central America	4.1%
South America	4.0%
Africa	2.5%

Marketing Tourist Facilities

People travel in order to satisfy a variety of needs and wants. The tourist industry has responded to these different needs and wants by developing a range of facilities, all targeted at slightly different groups of people. Hotels are a good example of this market positioning. In any major tourist area, hotels range from the basic "no frills" accommodations (small, modest rooms with few services) to luxury suites (large rooms, with many services available to patrons). Different hotel chains create different images in potential guests' minds through advertising, use of signs, locations, and so on. Each image is designed to appeal to a specific segment of tourists.

Tourist regions use the same approach to attract vacationers. They develop advertising and promotion campaigns that are targeted at specific markets — groups of people who will find the facilities of the region appealing. If a region has plenty of beaches and nightclubs, it will direct its messages to younger tourists; a mountainous region will market its physical setting with promotions aimed at hikers and campers; large cities target tourists who will enjoy theatre, clubs, and shopping. Through promotion and advertising, the tourist industries try to create a distinct, positive image of their regions. Magazine, newspaper, and television ads are popular forms of promotion for the industry.

In order to be effective, advertisers must understand what appeals to tourists. Researchers have concluded that all travellers

Figure 2.10 *A Typical Magazine Advertisement*
What might be the characteristics of people who would find this ad appealing?

are looking for value for their money, a variety of activities, friendly people, and good service. Specific markets look for other qualities as well. For example, people who vacation in cities want:

- shopping;
- a cosmopolitan atmosphere;
- cultural activities.

Those who book into resorts are looking for:

- night life and entertainment;
- luxury hotels;
- high-quality restaurants.

Outdoor enthusiasts will want:

- wilderness;
- lakes and streams;
- seaside facilities.

The goal of advertising is to inform specific target markets about the facilities that will appeal to them at the various destinations.

CHECKBACK AND APPLY

10. a) For each of the following three tourist destinations, write a list of ten words or images that come to mind.

- Acapulco
- Banff
- London, England

b) On what did you base your answer to a)?

c) How accurate do you suppose your images are of these places? Explain your answer.

11. What appeals to you as a tourist? List five places where you would like to vacation. What activities would you like to participate in while on vacation?

12. Design an advertisement to promote one of the tourist facilities in your area. Decide what segment of the market you will target, the image you should try to create, and the best method for communicating with the potential market.

Recent Trends in Travel and Tourism

The tourist industry has to respond to changing conditions both at home and around the world. For example, the traditional two-week vacation has almost disappeared as the only travel that a family or individual will do during the year. Quick "get-away" holidays are gaining in popularity as people with rushed schedules choose weekend trips to destinations such as the Bahamas, Las Vegas, or the Caribbean.

The travelling population is older now than in the past. People today live longer, healthier lives. Retired people who have substantial discretionary incomes expect to be able to travel, and the tourist industry is adapting by providing services that seniors find appealing. These services include organized tours, accommodations with housekeeping (cooking and eating) facilities, and more slowly paced recreational activities.

Transportation choices have changed in recent decades. An increasing share of travel has gone to the airlines, with less people travelling by trains and buses. Three factors are important when making transportation choices: comfort, convenience, and cost. If the differences in prices are not too great, many travellers will choose convenience and comfort over costs. The airline industry is highly competitive, which has kept seat prices down. In addition, airlines offer cut-rate prices at non-peak times, and "no-frills" charter packages, which have further attracted customers.

The airlines have made air travel more convenient by using **travel hubs**. The costs of flying international flights into every city with an airport would be prohibitive. Instead, smaller cities are connected by commuter routes to larger, regional centres. The international airlines fly into these regional centres, and passengers book connecting flights to smaller places. Figure 2.11 shows the travel hub at Frankfurt, Germany.

Figure 2.11 *The Travel Hub at Frankfurt, Germany*
In what ways do travel hubs make air travel more convenient?

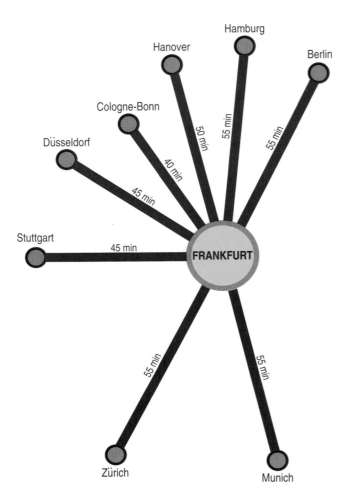

Tourism is a service industry. If it fails to change to meet the demands of the travelling public, it will not grow. The industry must also deal with some important issues in the coming years, including environmental concerns and cultural issues. The industry's success will depend on its abilities to deal with problems and issues, and to take advantage of new opportunities.

CHECKBACK AND APPLY

13. **a)** List advantages and disadvantages of air, rail, and bus travel. Use a chart to organize your ideas.

    b) Comfort, convenience, and costs are three important factors when choosing a mode of transportation. Which of these factors would be most important to each of the following groups of people? Why?

    - young people touring Europe
    - business people travelling to Asia
    - retired seniors escaping northern winters
    - families with young children on an annual vacation

14. Imagine that you were setting up a travel agency that specializes in serving clients over 55 years of age.

    a) What types of vacation packages would you develop to appeal to this group of people?

    b) What media would you use to advertise to this group? Give reasons for your choices.

15. Think about the changes and trends that you have observed in society and consider how these will affect the tourist industry. Predict how each of the following vacation options will fare in the next decade, giving reasons for your predictions.

    - two-week mid-winter vacations
    - three-day "get-away" trips
    - extended motor home vacations
    - cruises to tropical areas
    - safaris in game reserves

Case Study

Jet Lag

The world has been said to be "shrinking", not because it has changed in size, but because people are able to travel great distances more quickly. As more people travel more often, the importance of **time zones** around the world increases. The map on pages A2-A3 shows that the world is divided into 24 time zones of roughly equal size. In general, each time zone is one hour different from the zone next to it.

Most international travel is done by airplane, especially jets. Air travel is fast and jets can carry large numbers of people long distances. There are, however, some side effects of such rapid, long-distance travel. One of these is **jet lag** — the condition your body experiences after quickly crossing time zones. Jet lag includes headaches, fatigue, nausea, and disorientation. There are several methods of reducing the impact of jet lag. These include being well rested before setting out on a trip and drinking fruit juices instead of alcohol and caffeine. It might also help to begin thinking in terms of the time at your destination before leaving home. You might even want to set your watch to your destination's local time before take-off.

CHECKBACK AND APPLY

16. Refer to pages A2-A3 to help you answer this question.

    a) If the time in Ottawa is 8 a. m. on Monday, August 6, what date and time is it in:

    - Vancouver?
    - Beijing, China?
    - Moscow, U.S.S.R.?
    - Nairobi, Kenya?
    - Bangkok, Thailand?
    - Tokyo, Japan?

    b) How many hours, and in which direction, would you have to change your watch if you travelled from Toronto to Los Angeles?

17. **a)** Suppose you were a business person who had to make frequent flights to other continents. How might jet lag affect your performance?

    b) What groups of people will likely be most affected by jet lag? Why?

Chapter Review

- Tourism has grown rapidly in the last 40 years and is now a major industry in many countries.
- The tourist industry is made up of a number of different, interrelated businesses.
- Advertising and promotion to selected audiences are tools used to encourage people to choose specific destinations.
- Tourism affects the countries in which it occurs both positively and negatively. One of the most important consequences is cultural change.
- The values and needs of the population the tourist industry serves are changing. The industry must develop in new ways to continue to meet the demands of travellers.

Vocabulary Review

accessibility
discretionary income
jet lag
ripple effect
time zone
travel hub

Thinking About ...
TRAVEL AND TOURISM

1. **a)** Give five reasons to explain the rapid increase in tourism over the past few decades.

   b) What trends do you predict for the travel industry in the next 20 years? Give reasons for your predictions.

2. **a)** In general, what effects does tourism have on countries that are tourist destinations?

   b) Compare the effects tourism will have on a European country, such as Spain, and on a developing country — Cuba, for example. Organize your answer in a chart form.

3. Examine the information in Figure 2.12.

Figure 2.12 *Travel Destinations of Canadians*

Destinations	1980	1988
United States	89.7%	86.5%
Europe	5.8%	7.3%
Caribbean	2.3%	2.5%
Central America	0.9%	1.3%
Other	1.3%	2.4%
Total visits (thousands)	27 402	35 076

Source: World Tourism Organization

a) Suggest reasons why the United States is by far the most popular destination for Canadian tourists.

b) What trends are apparent in the data? Suggest reasons for these trends.

c) In what three ways could the Caribbean region or countries of Central America attract more Canadian tourists? Explain your ideas.

4. **a)** In what ways might a typical tourist cause cultural conflicts in a tourist destination? Some topics you might think about are:

   - dress codes
   - religious practices
   - behaviours in public
   - food preferences
   - buying habits

   b) What are some actions that tourists could take to minimize their negative effects on the local cultures?

LOOKING BACK AT TRAVEL AND TOURISM

 ## Atlas Activities

1. Refer to the time zone map on pages A2-A3. What reasons can you suggest for having different time zones around the world?

2. **a)** In the middle of the map on pages A2-A3 there is a line called the International Date Line. Examine it carefully. What occurs at that line?

   b) What happens to both the time and the date as you travel across the International Date Line from east to west?

   c) What importance would this line have for tourists travelling across the Pacific Ocean?

3. Using the same time zone map, answer the following questions.

   a) What is the approximate distance between Toronto and the following places?
   - Japan
   - Florida
   - the Netherlands

   b) Assuming that a jet travels at 900 km/h, how long would it take to travel from Toronto to each of the destinations listed?

   c) If the time in Toronto is 2 p. m., what is the time in Japan, Florida, and the Netherlands?

 ## Further Explorations

1. Obtain several editions of a newspaper that contains a travel section with travel advertisements. You should have at least seven travel sections.

   a) Label several sheets of paper, each with a different area or country of the world, such as the United States, the Caribbean, South America, Africa, Eastern Europe, Western Europe, the Middle East, South Asia, the Far East, and Australasia.

   b) For which area or country do you see the most articles and advertisements? Is the area or country one of the major destinations of Canadians as shown in Figure 2.12 (page 24)?

   c) Briefly describe the types of travel packages that are offered for each place, with their costs.

   d) Which areas or countries are the least expensive to travel to, from the area in which you live?

   e) Would the results of your research have been different at another season of the year? Explain your answer.

2. Use your school library to help you with this question. Select one country in the world, other than Canada, and research it in some detail.

   a) What types of transportation are found in this country? Which is the most important one?

   b) As a tourist, how would you arrive in this country? Where would you be likely to arrive?

   c) How would you travel around inside this country?

   d) What are some of the major problems this country faces with respect to its transportation system?

   e) What improvements could be made to the transportation system that would benefit tourists?

   f) What types of tourist resources or destinations does this country have?

3

Florida

AN ECONOMIC REGION

Figure 3.a   *A Space Shuttle Ready for Launch*

Figure 3.b   *A Tourist Resort at Clearwater Beach*

Figure 3.c   *Typical Strip Development*

Figure 3.d *Urban Development on a Mangrove Swamp, Madeira Beach*

Figure 3.e *Cypress Gardens: A Popular Tourist Destination*

STARTING OUT

1. Describe the location of Florida in relation to:
   - the rest of the United States;
   - Central America and the Caribbean Islands;
   - your home.
2. List seven economic activities that are shown in the photos. Approximately what proportion of the activities are related to tourism?
3. List three things that attract tourists to Florida.
4. From what you already know about tourism in Florida, what problems do you think the tourist industry is facing?

LEARNING DESTINATIONS

By the end of this chapter, you should be able to:
- understand some of the major characteristics of an economic region;
- describe Florida's development as an economic region;
- account for the rapid growth of the tourist industry in Florida;
- identify some issues facing Florida and understand the effects these may have on Florida's economy and on tourism.

Introduction

A homogeneous region is an area with one, or many, characteristics that set it apart from other areas. We might choose to define a homogeneous region by its:

- physical characteristics (such as climate or landforms);
- identifiable cultural patterns (such as language or history);
- common social qualities (such as its demography or lifestyles); or
- economic activities (such as development of natural resources or manufacturing).

In this study, we define Florida as a homogeneous region by focusing on the economic activities that are important throughout the region. Tourism is the dominant activity in Florida, our **economic region**.

Florida has developed a thriving tourist industry in the last few decades. It is the destination for almost 40 million tourists a year, most of whom come from the eastern part of the United States and Canada. Yet, at one point, Florida was considered to have little potential.

Geologically, the Florida Peninsula is relatively young. Long after the bulk of the continent was formed, the Florida Peninsula emerged from the oceans as part of the Atlantic coastal plain that stretches down the eastern seaboard and around the Gulf of Mexico. Barely above sea level, the Peninsula has a vast network of swamps, rivers, and lakes. Its coastline is dotted with many beaches which, combined with a hospitable climate, attract North Americans from harsher climates in other parts of the continent. These sun-seekers laid the foundation for the tourist industry in Florida.

Figure 3.1 *A Cypress Swamp*
Florida's location and landforms present both opportunities and problems. In what ways might they be considered resources?

CHECKBACK AND APPLY

1. In a group, list economic activities that you associate with Florida. Group the activities into categories, one of which will be tourism. What three categories do you think are most important to Florida's economy?

2. Suggest at least three characteristics that can be used to identify an area as an economic region.

3. What other places in the world could be designated as economic regions based on tourism? Name at least three places.

The Economy of Florida

During the 1980s, Florida had one of the fastest growing economies in the USA. In the first five years of that decade, for example, the number of jobs rose by 24 percent, an annual growth rate of almost 5 percent. Much of this growth was stimulated by the spending of tourists and retired people.

The service industries have benefited most from tourism in Florida. In total, service industries account for about four-fifths of Florida's **gross state product (GSP)**, the total value of all goods and services that are produced annually in a state. Figure 3.2 compares the relative importance of the main sectors of the economy.

Figure 3.2 *Economic Activities in Florida (1986)*

	Number of Workers	Percent of Florida's GSP*	Percent of U.S.'s GDP*
Agriculture	102 400	2	4
Mining	9 300	1	2
Manufacturing	517 200	11	19
Construction	339 500	7	5
Transportation, communications, and utilities	247 400	9	9
Wholesale and retail trade	1 238 800	19	16
Finance, insurance, and real estate	339 700	19	17
Community, social, and personal services	1 205 600	20	18
Government	701 900	12	10

*Gross State Product *Gross Domestic Product   Source: Florida Tourist Bureau

Tourists intent on enjoying themselves are willing consumers of services. They buy food, clothing, and souvenirs. They go to shows and concerts, rent cars, take tours, and use health-care facilities. The also pay for accommodations. Many real estate companies have become successful developing vacation resorts and retirement communities. Spending by American tourists alone in Florida in 1985 injected US$18 642 million into the economy.

Figure 3.3 *Tourist Expenditures by Americans, Top Ten States (1985)*

State	Expenditure (US$ millions)
California	32 507
Florida	18 642
New York	16 538
Texas	15 685
New Jersey	12 933
Pennsylvania	9 321
Illinois	8 662
Michigan	7 172
Nevada	6 912
Ohio	6 849

Source: U.S. Travel Data Center

Six of the top ten states for tourism are in the northern part of the USA. What besides weather would attract tourists to Florida?

The service sector of the economy is not the only sector to benefit from tourism. Other sectors also benefit. For example, the demand for resort facilities stimulates growth in the construction industry. The communications sector benefits by keeping vacationers linked to their home communities. This stimulation of activity on many fronts is called by economists a **multiplier effect**. Every dollar spent by tourists creates a much greater economic value in the region as it circulates among the various industries.

What stimulated this growth in the tourist industry in Florida? Before World War II, Florida was sparsely settled; most of the state was flat marshland or pine forest with few farms or roads in the interior. It was considered too hot and swampy for significant economic activity. This began to change in the 1950s with the development of some new technologies. Heavy earth-moving equipment that made canal building easier and faster became available. This meant that swampland could be drained and used for agriculture and urban development. A complex series of canals was constructed to drain the major swamp areas of southern Florida.

Large cities began to grow at key points in the state. Orlando, the site of Disney World, is located midway between the Gulf of Mexico and the Atlantic Ocean, on land once considered uninhabitable.

Another factor that helped to change the face of Florida was the introduction of low-cost, efficient air conditioners. It was now possible to escape the intense southern heat. Florida began to attract a large number of residents and tourists.

Figure 3.4 *Miami Beach*
How have economic activities taken advantage of the natural attractions of Miami Beach?

CHECKBACK AND APPLY

4. In what ways is technological change responsible for the development of Florida's tourist industry?

5. Refer to Figure 3.2. Which economic activities do you think would benefit least from tourism in Florida? Why?

6. Create a diagram that shows how the multiplier effect works. Include at least four stages in your diagram.

7. **a)** Figure 3.3 suggests that not all Americans are sun-seekers. What tourist activities do you think might attract people to New York? to Pennsylvania? to Michigan?

   b) In what ways will the economic impact of the tourist industry in these places be similar to that in Florida?

8. Florida has acted like a magnet in attracting tourists from the northern states and Canada during the winter months. It has also proved to be attractive to those who live in the Caribbean and in Central and South America.

   a) Refer to pages A28 and A29. Describe the location of Florida relative to the countries of the Caribbean.

   b) Suggest four attractions that Florida might have for people from these parts of the world.

The Cities of Florida

Much of the tourist development in Florida is centred on the cities, particularly Miami, Orlando, Tampa, and Jacksonville. Each has its own special brand of tourism.

MIAMI

The focus for tourists in Miami is Miami Beach, a 19.4 km island separated from the rest of the city by Biscayne Bay. Although Miami Beach has a permanent population of only 100 000, it can accommodate three times that number of tourists. This slender strip of land has over 300 hotels and plays host to 5 million visitors a year. Because it faces the Atlantic Ocean, the island is in danger of being eroded by waves and ocean currents. Over the years, breakwaters and reefs have been built to reduce the impact of these natural processes. This protects the investment of tourist operators.

On the southwest side of Miami is Little Havana, a community made up mostly of Cubans who fled their home country. Little Havana adds a distinctive cultural flavour to the city.

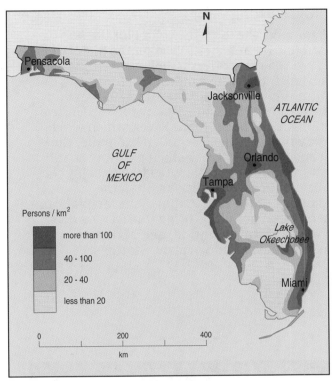

Figure 3.5 *Population Density in Florida*
What factors do you think helped to create this pattern of population density in Florida?

ORLANDO

Orlando is situated in the interior of the state in an area that boasts many lakes. Within the boundary of the city alone there are over 50 lakes. A number of tourist attractions, such as lake cruises and canoe routes, have taken advantage of this resource. But the most remarkable tourist attraction in the area is Walt Disney World. In the mid-1960s, the announcement that Walt Disney World was to be built stimulated the construction of facilities in Orlando to serve the millions of visitors who descend on it each year. There are, for instance, more than 3000 eating places in this city of 150 000 people. Other types of attractions — **theme parks** (parks built around a particular topic), entertainment complexes, and gardens — developed to take advantage of the huge flood of visitors to the area, making it a vast complex of tourist attractions.

TAMPA

Kilometre after kilometre of soft sand draws tourists to this part of the Florida Peninsula, but its biggest attraction is Busch Gardens. This African theme park is home to over 3500 animals, making it one of the world's largest zoos. Millions of tourists visit it annually. However, Tampa is not content to be only the home of the Busch Gardens. The city also has one of the nation's largest convention centres. It has encouraged tourism through the development of such attractions as a restored historic city centre, a pedestrian shopping mall, and a performing arts centre. In addition, docking facilities for cruise ships are planned for the coming years.

JACKSONVILLE

Jacksonville, the largest city in Florida, has built a tourist industry based on the history of the area. The city offers a replica of a 1564 French fort, the state's oldest existing plantation (which dates back to 1792), and, nearby, the oldest fishing village in the United States. Along the St. Johns River, which winds its way through the heart of Jacksonville, the city has built the Riverwalk, a pathway that allows visitors and residents alike to enjoy the beauty of the river while sampling the attractions of the city. Annual jazz festivals and other special activities bring tourists, as does the Mayport Naval Station, one of the east coast's largest aircraft-carrier stations.

The Physical Environment: A Basis for Tourism

Florida's rise to prominence as an economic region founded on tourism is largely due to its climate and beaches. As one Canadian vacationer remarked:

> 66 *When I retired as a police officer, I decided I was tired of winter and snow. Southern Florida has always had an attraction for me because the winters are warm and you don't have to put up with the cold weather and all the extra clothes. I like to golf, and even in the middle of January the weather is perfect for nine holes. From November to April, Florida sure beats Canadian winters!* 99

Hugh Ross

Winter temperatures across the state vary considerably, but summers are uniformly hot.

CHECKBACK AND APPLY

9. **a)** Compare the tourist industries in Miami, Orlando, Tampa, and Jacksonville. Some topics you should consider are:
   - location in the state
   - focus for tourist activities
   - importance of the physical environment
   - problems that may hinder tourism

   b) Which tourist facilities within the four cities interest you the most? Give reasons for your answer.

10. What role should the government of Florida play in supporting tourism in the state? Suggest four ways in which the government should be involved, offering reasons to explain your points.

11. Suppose you were an investor seeking to develop a new tourist facility. Where in Florida would be the most suitable place to locate each of these facilities?

    a) a high-priced beach resort for singles

    b) a theme park based on historic events

    c) a tour service focused on the wildlife unique to Florida

    d) a budget motel geared to families

    Explain why you made your decision for each facility.

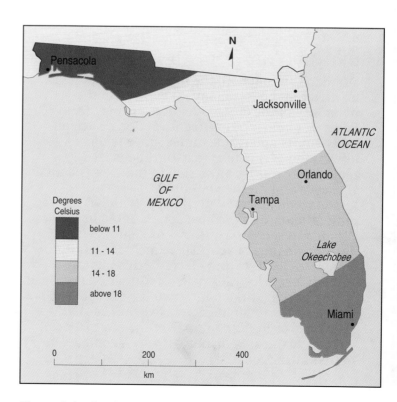

Figure 3.6a *Average January Temperatures*

Figure 3.7 *Climate Statistics for Pensacola, Tampa, and Miami*
The top row for each city shows average temperatures in degrees Celsius.
The bottom row shows average precipitation in millimetres.

Cities		J	F	M	A	M	J	J	A	S	O	N	D
							Months						
Pensacola	Temp. (°C)	16	17	19	23	27	29	31	31	29	25	20	16
	Precip. (mm)	8	9	12	16	19	23	24	24	22	17	12	8
Tampa	Temp. (°C)	21	22	24	27	30	32	32	32	31	28	24	22
	Precip. (mm)	11	12	14	17	20	22	23	23	22	19	14	12
Miami	Temp. (°C)	23	24	25	26	28	29	31	31	30	28	26	24
	Precip. (mm)	17	17	18	20	22	24	24	25	24	22	19	17

Source: Florida Tourist Bureau

How would you describe the climates of Pensacola, Tampa, and Miami?

CHECKBACK AND APPLY

12. Construct a climate graph for each of the cities named in Figure 3.7.

13. a) Compare the climates of Pensacola and Miami by noting for each place:
- warmest monthly temperature;
- coolest monthly temperature;
- temperature range;
- annual precipitation.

b) What reason can you suggest to explain the variations between these two places?

c) During which season(s) will Pensacola attract the most tourists? Why?

14. a) Compare the temperatures of Tampa and Miami. When do the greatest differences in temperature between these two cities occur?

b) In what ways would the differences in temperature affect the tourist industries in these cities?

15. a) What other types of economic activities, besides tourism, would benefit from Florida's climate?

b) What effect will hurricanes, which occur most frequently in the fall of the year, have on economic activities in the state?

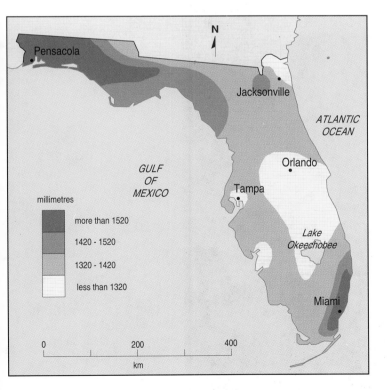

Figure 3.6b *Average Yearly Precipitation*
What effect do water bodies have on the climate of Florida?

Famous Tourist Destinations

Walt Disney World

In the early 1960s, much of the land around Orlando was untouched wilderness, with just a few cattle ranches and orange groves. The swampland was inexpensive, but there were few buyers. Orlando itself was a small, slow-paced city. At that time, however, the Disney Corporation was quietly on the lookout for land to purchase for an exciting new enterprise. Who could have guessed that this land, once purchased, would be the site of the largest and one of the most profitable theme parks in the world?

In October, 1965, Walt Disney purchased 10 977 ha of land near Orlando for Disney World. It opened officially in October, 1971. In its first year of operation, almost 11 million people visited the park.

Walt Disney originally envisaged a futuristic city, called EPCOT (Experimental Prototype Community of Tomorrow), where people would actually live. The nature of EPCOT has changed somewhat, but it has been built as part of the Disney World complex. In the late 1980s, the Disney Corporation created the Disney-MGM Studio Park, adding another part to the complex.

The concept behind Disney World, as with other similar theme parks, is to transport visitors into another world where they can forget about everyday problems and simply enjoy themselves. The more they enjoy themselves, the more willing they are to spend money.

An Early Photo of Walt Disney with Mickey Mouse

Figure 3.8 Disney World Disney World has many attractions, such as rides, permanent exhibits, and recreational areas.

The Disney Corporation is financially successful because it puts the needs of its customers first, and does so with excellence. Behind the scenes, the work of painters, electricians, cleaning crews, technicians, and carpenters keeps the park looking impeccable.

Disney World and the EPCOT Center now attract close to 25 million visitors a year — equal to the entire population of Canada.

Unique features of Disney World include:

- environmentally safe garbage and sewage treatment and recycling systems. The garbage disposal system uses a vacuum system and central collection depot;
- a wilderness area which makes up one-quarter of the site;
- huge underground service systems that allow workers to walk to any site in the park unseen by park visitors and that provide electricity and water throughout the complex;
- a monorail system linking hotels and recreation areas.

Figure 3.9 *Orlando and Its Environs*
Disney World is centrally located in Florida and is well served by highways. Note that there are a number of other major tourist attractions in the same general area.

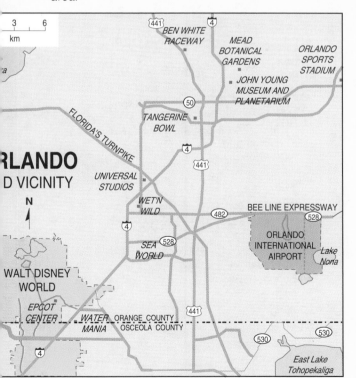

CHECKBACK AND APPLY

16. What attractions does Disney World offer to tourists? Which of these attractions appeal to you? Why?

17. Would the development of other theme parks and tourist attractions in the Orlando area help or hurt business at Disney World? Explain your reasoning.

18. Although Florida itself is considered a region, there are smaller sub-regions within it. What characteristics could be used to define the area around Orlando as an economic region? Explain.

19. The Disney Corporation's impact extends far beyond its theme parks. It has created products that sell around the world. It has also come to represent the United States through its image and corporate values.

    a) Describe the image that the Disney Corporation has in the minds of consumers.

    b) Name 10 products that the Disney Corporation produces or that display a Disney character.

    c) How do these products help to promote the theme parks the Disney Corporation operates?

Other Economic Activities in Florida

Economic growth in Florida is not limited to tourism. The benevolent climate has given rise to a thriving citrus industry. Oranges are Florida's single most important agricultural product. Grapefruits, limes, tangerines, and tangelos are also grown. In total, this state produces about two-thirds of the nation's citrus crops. Most of the citrus groves are in south-central Florida, between Orlando and Lake Okeechobee.

Manufacturing has also made real gains in the state in recent years, generating thousands of new jobs. There is a concentration of communications equipment manufacturing in the Fort Lauderdale and Tampa area. Much of this equipment is destined for military and space programs. Manufacturers have been attracted by the warm climate, low labour costs, and inexpensive land. Industrial parks have been developed to house the new industries in places such as Jacksonville, St. Petersburg, and Miami.

Figure 3.10 *An Orange Grove in Florida*

CHECKBACK AND APPLY

20. a) In what forms are consumers able to purchase Florida orange juice?

b) What types of product would compete with Florida juice for the consumer market?

c) What advantages does Florida juice have in the North American market?

21. Suppose you are a business person trying to decide where to locate a manufacturing operation. What advantages and disadvantages does Florida have? Record your ideas in a chart.

22. Draw a map showing the location of the major citrus groves in Florida.

Florida
THE REGION AT A GLANCE

Area	151 939 km²
Population (1988)	12 335 000
Population density	81.2 persons/km²
Urban population	91%
GSP* per capita (1988)	US$16 546
State parks and recreation areas	
—number of sites	129
—land area	112 312 ha
—number of visitors (1987)	14 290 000
Tidal shoreline	
—Atlantic	5330 km
—Gulf Coast	8152 km
Rate of population growth	2% per year
Population	
—aged 65 and over	18%
—national average	12%
Annual death rate	
—Florida (1987)	10.7/1000 people
—national average	8.7/1000 people
Main industries	tourism, manufacturing, agriculture

*Gross State Product

Florida's Tourist Industry

Winter is the peak season for Florida's tourist industry, particularly in the southern part of the state. Tourism also increases during the winter school break and other major holiday periods throughout the year. During the **shoulder seasons** — those nontraditional travel periods — rates are lowest and facilities least crowded.

Most tourists travel to Florida by car. This is the least expensive method of transportation for families and small groups of people and it provides a high degree of personal freedom. There are three major highway routes into the state, **Interstate Highways** #75, #10, and #95. These are limited access highways designed for long-distance travel within and between states. Their construction facilitated the movement of people into the southern part of the United States and stimulated the tourist industry.

More than half of all North Americans live in the eastern part of the continent, within two days driving time of Florida. A variety of tourist facilities have developed along the Interstate Highways to serve travellers, particularly at interchanges. Restaurants, service stations, resorts, and shopping plazas are some of the tourist-oriented facilities located on major transportation routes. There are over 150 000 campsites in 1000 campgrounds in the state, most of which are open year-round.

Motels and hotels also benefit from the high volume of traffic. In recent years, hotel chains have become very successful at servicing Florida's tourists. Holiday Inn, Days Inn, and Motel 6 are some of the most conspicuous chains. These chains provide **toll-free numbers** so that clients can call free of charge to arrange their accommodations. Most hotel chains provide a standardized set of facilities, such as swimming pools, restaurants, and room features.

Florida has pioneered other types of tourist-oriented businesses. For example, its car rental business, which serves vacationers who arrive by airplane, has become one of the largest of any state in the U.S. Computerized reservation systems have allowed this business to rapidly improve efficiency and better meet the needs of customers. Intense competition among large rental chains has kept rates low and volume high.

Figure 3.11 *Highway Development in Florida*
The Interstate Highway system has been a major factor in bringing tourists into Florida. Motels often locate at the intersections of the Interstate to serve tourists travelling by car.

23. Turn to the map on pages A26 and A27.

a) From which parts of North America do you think most tourists would come? Why?

b) Examine Figure 3.12, which shows the origins of tourists travelling to Florida by car. Mark each of the places listed on an outline map of North America. Give your map an appropriate title.

Figure 3.12 *The Top Ten Origins of Tourists to Florida, by Car*

1. Georgia	6. Texas
2. Ohio	7. Alabama
3. New York	8. Louisiana
4. Tennessee	9. Ontario
5. Michigan	10. Pennsylvania

Source: U.S. Travel Data Center

c) What pattern does your map show about the origins of tourists to Florida?

d) Suggest two implications that this pattern has for advertising Florida's attractions.

e) Suggest three reasons, other than for advertising, why the tourist industry in Florida is interested in getting and using statistics of this kind.

24. Most tourists to Florida arrive by car or by airplane. List the advantages and disadvantages to a tourist of each of these modes of transportation.

25. Ninety-two percent of all visitors to Florida are repeat visitors. Why do you think this is so?

26. Many of the tourists to Florida come in family groups. List some of the actions you would take to attract and accommodate families if you were:

a) a motel owner

b) a restaurant operator

c) a theme park manager

d) a tour bus booking agent

27. Figure 3.13 lists Florida's top ten tourist attractions. Locate these tourist sites on a map of Florida.

Figure 3.13 *Florida's Top Ten Tourist Attractions*

1. Walt Disney World (Orlando)
2. EPCOT Center (Orlando)
3. Disney-MGM Studios (Orlando)
4. Sea World (Orlando)
5. Spaceport USA (Cape Canaveral)
6. Busch Gardens (Tampa)
7. Cypress Gardens (Winter Haven)
8. Florida's Silver Springs (Silver Springs)
9. National and State Parks and Reserves (throughout state)
10. Miracle Strip Amusement Park (Panama City)

Source: Florida Dept. of Commerce, Division of Tourism

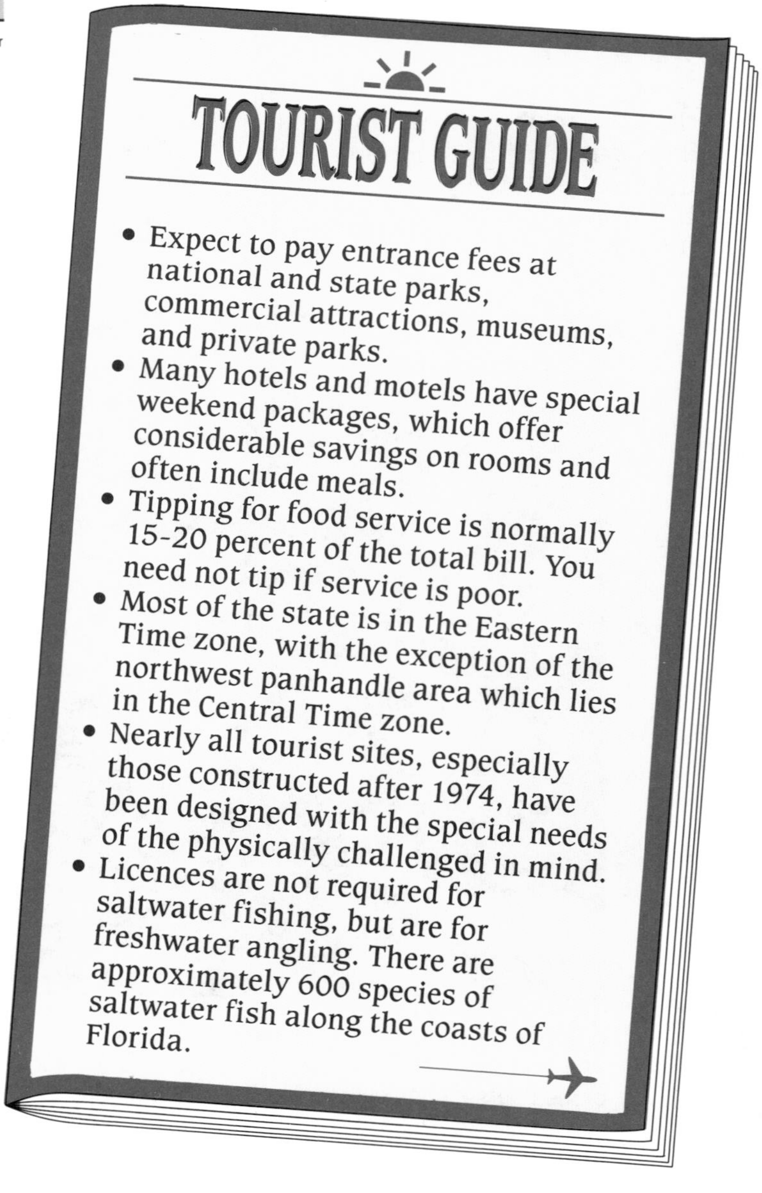

TOURIST GUIDE

- Expect to pay entrance fees at national and state parks, commercial attractions, museums, and private parks.
- Many hotels and motels have special weekend packages, which offer considerable savings on rooms and often include meals.
- Tipping for food service is normally 15-20 percent of the total bill. You need not tip if service is poor.
- Most of the state is in the Eastern Time zone, with the exception of the northwest panhandle area which lies in the Central Time zone.
- Nearly all tourist sites, especially those constructed after 1974, have been designed with the special needs of the physically challenged in mind.
- Licences are not required for saltwater fishing, but are for freshwater angling. There are approximately 600 species of saltwater fish along the coasts of Florida.

Challenges Facing the Tourist Industry

Despite the many attractive features that Florida has for tourists, a number of issues cloud the future of the state. One issue is the destruction of the natural environment. As Florida has grown and developed, the natural environments of the state have changed considerably. Swamplands have been drained, rivers rerouted, and forests chopped down. The **Everglades**, for example, have been drastically affected. The Everglades are an area of natural swampland in southern Florida that contain a rich variety of plant and animal life. Over the years, the Everglades depended on a continuous inflow of fresh water for life and vitality. Unfortunately, the construction of canals and the use of fresh water for irrigation and industry has cut the supply of water to the Everglades. Slowly, this area is ecologically dying. The Florida panther, for example, has lost most of its habitat and is near extinction. Similarly, the manatee — a large, gentle, freshwater mammal — has been killed in large numbers by human activity. There are programs underway, however, to save the panther, the manatee, and the Everglades. Several rivers have been returned to their original routes, wildlife sanctuaries have been established, and publicity campaigns have raised awareness of the need to preserve the environment. The success of these programs has still to be determined.

Ironically, southern Florida, which has been a swampland for centuries, now suffers from severe water shortages. Human activity has so altered the flow of surface and ground water that urban development will be restricted due to water shortages.

The coastline environments of Florida have also suffered. Most of Florida's development has taken place along the coastlines, first the Atlantic Coast and then the Gulf Coast. The coasts, however, are ecologically sensitive and can be easily damaged by careless human activity. In southern Florida, where winter temperatures are warmest, many of the coastal inlets are lined with **mangrove trees**. These short trees grow in shallow saltwater in tropical and subtropical regions. Their roots provide a habitat for numerous birds, fish, and animals, and protect the shoreline from erosion. Urban development usually results in the destruction of the mangrove swamps. Because of their value, laws have been enacted to protect mangrove trees.

Figure 3.14 *The Everglades*
The Everglades can be saved, but there is little time in which to do so.

A second challenge to the tourist industry in Florida is population density. The state's population growth rate is the highest in the United States. About 1000 people a day move to Florida, many in search of sunny climates, while others move there to take advantage of the strong economy and plentiful jobs. With this growth comes the need for more homes, roads, schools, sewage treatment facilities, water purification plants, and the like. Much of this urban development is in direct conflict with tourist activities based on natural attractions.

Immigration accounts for much of Florida's population growth. Florida has acted like a magnet for immigrants — both legal and illegal — from the Caribbean region and Central and South America. Thousands of immigrants, fleeing political and economic problems in their home countries, have arrived in Florida. This has changed the cultural face of the state, adding a Latin American aspect to the existing mix of Black, white, and native groups. Adjustments have not always been easy for either Floridians or the new Americans. Conflicts have arisen over language, politics, and economic development. There are signs, however, that these conflicts are being resolved.

The location of Florida at the southern tip of the United States has made it a logical entry point for illegal drugs from Latin America. The flow of cocaine, marijuana, hashish, and their bi-products, has been estimated to be in the billions of dollars. The consequences of this drug trade include an enormous toll in human lives as well as a high crime rate. The crime rate in the Miami area, for example, is among the highest in the country. During the early and mid-1980s, the crime problem became so great that it held back the development of the tourist trade. Despite some victories on the part of law enforcement agencies, and a revival of the tourist trade in the Miami-Fort Lauderdale area, the problems associated with the drug trade will be around for the foreseeable future.

CHECKBACK AND APPLY

28. Outline the ways in which each of the following has negative consequences for tourism in Florida.
    a) the destruction of the Everglades
    b) reduction of the mangrove swamps
    c) population growth
    d) the drug trade

29. a) Describe some tourist activities that could take place in an area of the Everglades that has been preserved in a natural state.
    b) List some types of activities that are incompatible with a natural area like the Everglades.

30. Suggest three laws that the government of Florida might pass to promote tourism without slowing the growth of the population. You might consider such ideas as:
    • controlling uses of the land
    • imposing building restrictions
    • identifying land with little tourism potential
    Think creatively about the use of laws and government regulations.

31. Florida has the highest proportion of senior citizens of any state in the U.S. Close to 18 percent of the population is aged 65 or over.
    a) Why do you think Florida has such a large number of senior citizens?
    b) List types of facilities or services the seniors of Florida will require. Think particularly of the special needs of that age group.

32. Global warming is a problem that the world has not yet faced to any real degree. Average temperatures are predicted to rise by several degrees over the coming decades, causing drought in some areas and melting portions of the world's ice caps. Scientists suggest that the higher sea levels will result in an increase of about 85 000 ha of wetlands in Florida. Brainstorm a list of consequences of global warming and rising sea levels for Florida and its tourist industry.

LOOKING BACK AT FLORIDA

Chapter Review

- The tourist industry can be used to define Florida as an economic region.

- The tourist industry is based on the attractive physical setting and the subtropical climate of the area.

- Huge investments have been made to provide facilities for tourists. These facilities provide jobs for the people of Florida.

- Walt Disney World is the largest and most well-known tourist attraction in Florida. It spurred the development of a huge number of other businesses geared to serve vacationers.

- Environmental problems related to the fragile ecosystems of the state have cast a cloud over the future of tourism.

- Population growth and the problems of large cities are other challenges with which the people of Florida are coming to grips.

Vocabulary Review

economic region
Everglades
gross state product (GSP)
Interstate Highway
mangrove tree
multiplier effect
shoulder season
theme park
toll-free number

Thinking About...

FLORIDA AS AN ECONOMIC REGION

1. Make a chart to show the advantages and disadvantages of Florida for tourism. Consider such things as: location, climate, tourist facilities, lifestyle.

2. Does Florida deserve the nickname "the sunshine state"? Use good arguments to support your point of view.

3. What evidence would you give to support the claim that Florida is a homogeneous economic region?

4. **a)** What do you suppose were the three most important reasons for choosing Orlando as the site for Walt Disney World? Explain your reasons.

   b) Name two other sites in North America that might have been considered for Walt Disney World. Why do you think they were not chosen?

5. **a)** Figure 3.15 shows the population growth of Florida over the past decades. Predict the population of Florida in 2010.

   b) In what ways might high population growth affect tourism? List and explain five ways.

Figure 3.15 *Population of Florida*

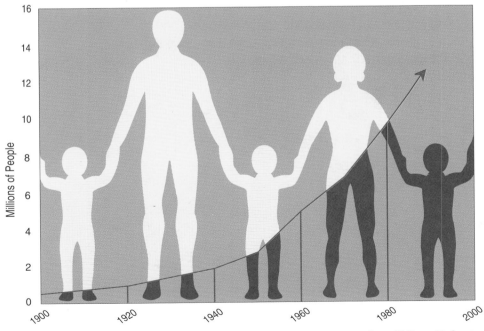

Source: U.S. Bureau of the Census

6. Describe the effect on Florida's tourist industry if each of the following were to happen.

    a) economic growth in California doubled

    b) hurricanes hit the area three years in a row

    c) U.S. officials destroyed the South American drug trade

7. Imagine you attended a Florida secondary school. What careers or job futures could you anticipate? What kinds of programs do you think might be available at colleges and universities in Florida?

8. Imagine that you are a developer interested in building a major tourist attraction in Florida.

    a) What type of tourist attraction would you develop? Give reasons for your choice.

    b) Who would be your target market (the group of people you hope to attract)? Give the characteristics of this group (for example, age and interests).

    c) Where in the state would you locate your attraction? Explain your answer.

    d) What specific features would your attraction have? Include an outline of the site showing its features.

    e) Describe the advertising campaign you would develop to attract tourists. Consider such things as: the medium or media you would use; where your ads would be displayed; what message your ads would convey.

    f) Outline some of the special needs you would have as you construct and operate your tourist attraction. You might list such things as highway access, labour needs, suppliers, and accommodations.

 Atlas Activities

1. Study the maps on pages A26 and A27.

    a) Describe the location of Florida relative to:
    • Canada and the United States
    • Mexico, Cuba, and the countries of the Caribbean

    b) Geographically, what does Florida have in common with Mexico and the countries of Central America?

    c) Based on the information given in these maps, select a location in North America that is as different as possible from Florida. Describe and account for the differences.

2. Using the information given in the maps on pages A26 and A27, write a short paragraph about Florida.

3. What characteristics of southern Florida make it unique in North America? Make specific reference to the maps on pages A26-A29.

4. Look at the world population map on page A4. Name six other areas in North America with a population density similar to that of Florida. What might account for the density of population in these areas?

5. Refer to the maps on enrolments in educational institutions on page A7. Compare the information given for Florida with that given for your area. Suggest reasons for similarities and/or differences you observe.

Further Explorations

1. Select a major tourist attraction in Florida and research the following:

   a) its location (include a map)

   b) its attractions

   c) the type of people it attracts

   d) how to get there

2. In a chart similar to the one that follows, compare the different types of citrus fruits grown in Florida.

Citrus Fruits	Main Growing Areas	Growing Conditions (soil, climate, etc.)
oranges		
grapefruits		
limes		
tangerines		
tangelos		

3. Select two major problems or issues facing Florida in the 1990s and beyond. For each problem or issue, explain:

   a) its nature;

   b) the part of the state most affected;

   c) who/what it affects most;

   d) its effect on tourism;

   e) possible solutions;

   f) new problems your solutions may cause.

4. Design and conduct a survey in your school to find out about people's experiences as tourists in Florida. As a group, brainstorm some topics that would be suitable to ask people about. You might include questions about time of visit, length of stay, accommodations used, and places visited. Copy your survey and complete at least 10 interviews with people who have vacationed in Florida. Analyze your data and present your results in a format of your choice. What conclusions can you draw from your survey?

5. Compare the tourist facilities in Florida to those in your own area. Use a chart to organize your points. Some topics you might want to consider are:

   • natural attractions
   • cultural and historic features
   • tourist facilities
   • problems that limit development
   • the future of the tourist industry

6. Investigate the employment opportunities available in a theme park. Which of these jobs appeals most to you for summer employment? Why?

4

HOLLYWOOD

Los Angeles

A CITY-CENTRED REGION

Figure 4.a   *Marina del Rey*

Figure 4.b   *Shopping Centre Rodeo Drive*

Figure 4.c   *Wholesale District in Los Angeles*

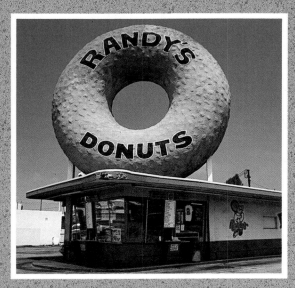

Figure 4.d *Donut Shop Near Los Angeles Airport*

Figure 4.e *Hills of Los Angeles*

STARTING OUT

1. What are some advantages of Los Angeles' location in terms of climate, tourism, and trade?
2. What can you tell about Los Angeles' physical geography from these photos?
3. What do you suppose attracts tourists to Los Angeles? Give at least four suggestions.
4. Based on what you see in the photos, name five jobs that you might do if you lived in Los Angeles.

LEARNING DESTINATIONS

By the end of this chapter, you should be able to:
- identify the characteristics that make Los Angeles a city-centred region;
- offer reasons to explain the rapid growth of the city;
- recognize the potential for tourism in Los Angeles;
- identify problems that arise because of Los Angeles' location and circumstances.

Introduction

Los Angeles is the second largest city in the United States. It is a major industrial and commercial centre whose very name conjures up images of movie stars, crowded beaches, and sunshine. In fact, there may be no other city in the world that represents glamour and wealth more than Los Angeles. The L.A. lifestyle has been characterized as informal, casual, friendly, fun-filled, and permissive. It is a diverse city, populated by people from almost every country of the world.

Los Angeles has not always been this way. One hundred years ago, it was an isolated little community far from the main population centres of the United States. Its early development was more directly influenced by Mexico, to the south, than by America. This influence has given Los Angeles a unique character and outlook on life. Today, although the Los Angeles area occupies only 5 percent of the state's land, it contains over half of California's population, 56 percent of its manufacturing output, and 60 percent of its wholesale trade sales.

Because of the growth that has taken place in the past decades, the city and its 80 or so surrounding communities have grown together. Each looks much like the others, so it is difficult to tell where one ends and another begins. The area has been described as "a hundred suburbs in search of a city". Collectively, these communities help to create a **city-centred region**.

1. Many of our images of Los Angeles come from movies and television programs.

   a) Brainstorm a list of words that describe your image of Los Angeles.

   b) What TV programs and/or movies help to create this image?

2. What factors do you suppose were important in causing Los Angeles to change from a small village to a major city? List three.

3. Describe in your own words what you think the term "city-centred region" means. Compare your definition with the information given in the next section of this chapter.

Los Angeles as a City-Centred Region

Los Angeles, one of the most famous cities in the world, can be considered to be the focus of a city-centred region. This is because the direct influence of the actual city of Los Angeles spreads well beyond the boundaries of the basin in which it lies. The city's influence extends up to 300 km away, both economically and culturally. It is the city of Los Angeles that imposes the character, business, and entertainment on the whole region.

The exact boundaries of the L.A. region are difficult to define. The region could be defined as the area in which L.A. newspapers are widely sold or in which L.A. television stations are regularly viewed. Alternatively, it could be defined as the area from which people commute to L.A. to work or to attend events, such as live theatre or sports. Using these criteria, cities such as San Diego, Palm Springs, and Santa Barbara would all be included within the Los Angeles city-centred region. To keep this study to a manageable size, this chapter will concentrate on the urban area within the Los Angeles "basin", an area of heavy urban development on plains and foothills which are ringed by mountains.

Figure 4.1 *The Los Angeles Basin*
The city of Los Angeles is one of a large number of cities located between the Coastal Mountains and the Pacific Ocean.

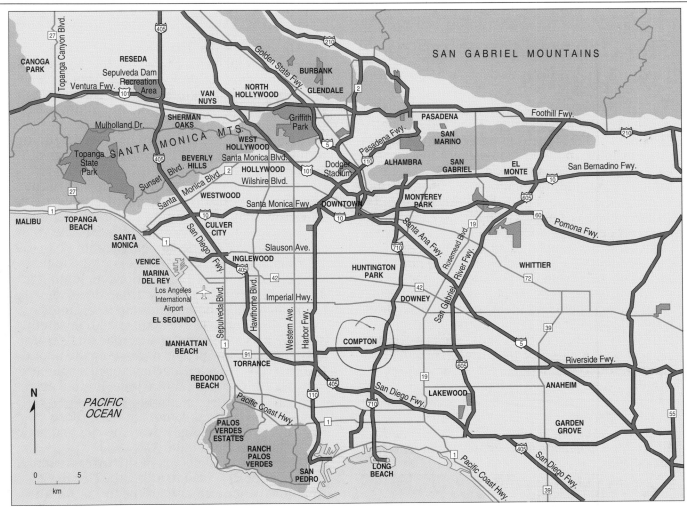

Figure 4.2 *The Los Angeles Metropolitan Area*
Find three things on this map that indicate that Los Angeles is a city-centred region.

The concerns in a city-centred region are different from those in other regions. Since much of the region is built up, the major issues relate to the problems of managing large numbers of people in a relatively small area. These issues include traffic congestion, pollution, crime, rapid population growth, loss of farmland, and concern about the quality of life.

Los Angeles has been one of the fastest growing cities in the United States. As it grew, nearby cities and towns, such as Hollywood, San Pedro, and Westwood, became part of the municipality of Los Angeles. This process, called **annexation**, makes providing services within an urban area simpler since control is held by one municipality. Outside of the actual city of Los Angeles, in Los Angeles County, there are 80 independent communities, including such well-known places as Beverly Hills, Pasadena, Santa Monica, Burbank, and Long Beach. While the city itself has a population of 3 million, there are 14 million people in the Los Angeles metropolitan area.

CHECKBACK AND APPLY

4. What five indicators might you use to find the extent of the influence of a city? (Two ways have been mentioned in this section of the chapter.)

5. Cities extend their influence beyond their own borders through the services they provide and the businesses they contain. If you live in a city, identify ways in which your city directly or indirectly influences the region around it. Consider such things as land values, transportation, and water sewage. If you do not live in a city, how is your area influenced by one or more cities? List at least four ways.

6. Describe five changes that take place in a city as its population increases. Identify both positive and negative changes.

The Urban Structure of Los Angeles

The Los Angeles area began, as many other urban areas have, with a collection of small, dispersed villages focused on agriculture and other primary industries. However, L.A.'s port gave it an extra advantage, and it grew faster than its surrounding communities. But, by the 1920s, the other communities — most located 8-24 km from Los Angeles — had developed their own city centres and were offering a wide range of services. Communities such as Pasadena, Santa Monica, and Long Beach began taking away business from Los Angeles and they grew rapidly as well. As these places expanded, the rural areas separating them were reduced, most noticeably along important transportation routes. By the end of the 1940s, the importance of the Los Angeles core had diminished in comparison to the importance of the peripheral communities.

Since the 1950s, the central part of Los Angeles has taken on a new life, largely because of new building technologies. Los Angeles is in an area that suffers from periodic earthquakes. Prior to the 1950s, the heights of buildings were restricted to make sure that they could withstand tremors. However, improved building techniques since the 1950s made buildings of 12 stories and more possible. This led to greater concentrations of offices and residences in the Los Angeles core. City officials responded to the new growth by encouraging cultural functions, such as Sunset Boulevard and the Miracle Mile. The city of Los Angeles quickly regained its role as a business and cultural leader of the region.

Because of height restrictions on buildings and the dispersed nature of urban development in the area, most homes are separate, single-family dwellings. These take up a huge amount of space and require extensive road and service systems. To allow people to move easily about the area, Los Angeles has developed a complex network of **freeways**, multi-lane, limited access highways designed to move large numbers of people at high speed. Unfortunately, these freeways are notoriously overcrowded and major traffic jams occur regularly.

Freeways are just part of the **infrastructure** that must be provided to allow the city to function. Other elements of the infrastructure include such things as: a reliable supply of energy, safe water supplies, efficient collection and treatment of sewage, and garbage collection and disposal. The dispersed nature of the L.A. area, and the region's rapid growth in recent years, has made expansion of these services difficult.

The growth rate of Los Angeles County is slowing down. The area has now been almost completely converted to urban uses and there is little land that can be developed without tearing down existing buildings. Since there is a high demand for the land, it is expensive to buy. Figure 4.5 shows that the periods of peak growth were in the early decades of this century. Growth has continued, but at a somewhat slower rate. Faster growth is occurring in the municipalities surrounding Los Angeles, where land is less expensive and the area less urbanized.

Figure 4.3 *Populations of Metropolitan Statistical Areas (1988)*

Los Angeles-Anaheim-Riverside	13 769 700
San Francisco-Oakland-San Jose	6 041 800
San Diego	2 370 400
Sacramento	1 385 200
Fresno	614 800
Bakersfield	520 000
Stockton	455 700
Salinas-Seaside-Monterey	348 800
Santa Barbara-Santa Maria-Lompoc	343 100
Modesto	341 000
Visalia-Tulare-Porterville	297 900

Source: U.S. Bureau of the Census

Metropolitan Statistical Areas include the main city plus other nearby communities with strong economic and social ties. What does this chart show about Los Angeles?

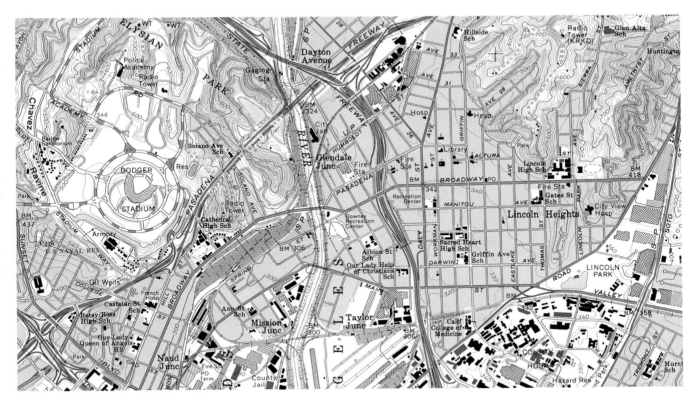

Figure 4.4 *Topographic Map of Los Angeles*
About what portion of the land in Los Angeles is occupied by roads? What other land uses can you identify in the topographic map?

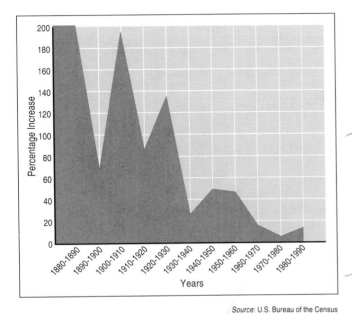

Source: U.S. Bureau of the Census

Figure 4.5 *Percentage Increase in Population, Los Angeles County*
Give one reason for each peak and valley shown in this graph.

CHECKBACK AND APPLY

7. a) Describe the urban development that has taken place in Los Angeles since the 1950s.

b) What three factors have contributed to this growth?

8. Explain why the Los Angeles area came to be called "a hundred suburbs in search of a city".

9. Make three sketches to illustrate the early, the pre-1950, and the present urban development of Los Angeles. Your sketches should show the size of communities, transportation connections, heights of buildings, and amounts of rural land.

10. a) Make a list of seven to ten services or goods that are normally found only in large cities like Los Angeles.

b) If Los Angeles had not developed a stronger role in the past four or five decades, would these services and goods be available in the area? Explain your answer.

Tourism Issue: Smog in Los Angeles

Los Angeles, as a large, fast-growing city, has its share of urban problems. Some of these problems have reached crisis proportions and must be tackled in the near future. Los Angeles is dominated by the car: it has built freeways to accommodate the millions of vehicles used there. Unfortunately, the population of the region has outgrown the ability of the freeways to handle the traffic. Traffic jams are among the largest in North America and can last for hours. At times, the traffic grinds to a halt and nothing moves. This condition is referred to as a **gridlock**. Gridlocks occur because there are too many cars for the roads. Most vehicles carry only one person and much of the freeway has not been repaired or updated in years.

Figure 4.7 *Traffic Problems in Los Angeles*
Traffic jams in Los Angeles have become part of the way of life for most people who live there. Wasted hours sitting in traffic on the freeway are a common frustration.

FRIDAY'S AIR QUALITY Hourly average pollution, 7 a.m. to 6 p.m.

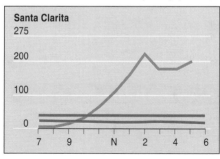
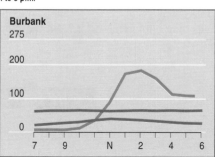
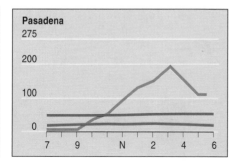
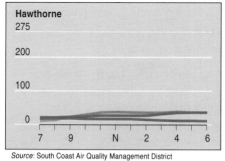
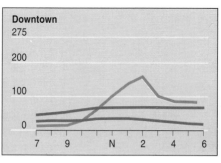
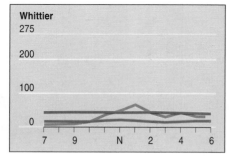

Source: South Coast Air Quality Management District

Figure 4.6 *Pollution in the Los Angeles Basin*
These graphs illustrate the daily pattern of dangerous air pollutants on a typical summer day at various locations in the Los Angeles basin. Think of one law you would pass that could reduce the smog without hurting L.A.'s economy.

Ozone: invisible, irritates, and impairs breathing
Nitrogen dioxide: brown, impairs breathing
Carbon monoxide: invisible, reduces blood's oxygen

Pollutant Standard Index (PSI)
Ozone, hourly average: carbon monoxide,
8-hour average: nitrogen dioxide, 24-hour average

0 - 50	good
51 - 100	moderate *(100 is federal standard)*
101 - 200	unhealthful *(200 is 1st stage episode)*
201 - 275	very unhealthful *(275 is 2nd stage episode)*
275+	hazardous

One of the consequences of this enormous use of cars is the air pollution that hangs over Los Angeles most of the year, and which is especially severe in the summer. Pollutants from motor vehicles react to sunlight and form photochemical oxides, more commonly referred to as **smog**. Smog is serious because it not only affects plants and animals, but also threatens human life. The smog in Los Angeles is the worst in the United States and continues to worsen. The air in Los Angeles County is considered unhealthy an average of 165 days in each year. Plans have been drafted to reduce the pollution released into the air. These plans include banning gas powered lawn mowers and electronically monitoring vehicles on the freeways. A great deal must still be done to improve the quality of air in Los Angeles.

CHECKBACK AND APPLY

11. Locate each community named in Figure 4.6. You can use the map on page 47 to help.

    a) Describe the trends in ozone pollution through a typical day and explain what might account for the highs and lows. (*Note*: When ozone is found high in the atmosphere, it protects us from harmful solar radiation. When ozone is concentrated at ground level, it is dangerous to health.)

    b) Which of the communities have the most severe air pollution? By examining Figure 4.6 and by considering the location of the place named, suggest reasons why these centres have such high levels of pollution.

Figure 4.8 *Cross-Section of the Los Angeles Basin*
This cross-section of the Los Angeles basin illustrates how the physical geography of the region has contributed to the air pollution problems by limiting the circulation of air necessary to cleanse the city air.

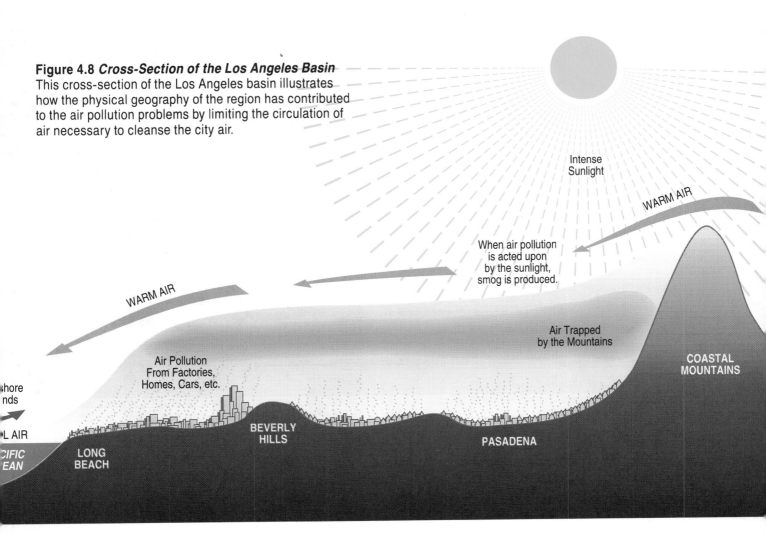

Los Angeles' Economic Structure

Los Angeles is the industrial centre of the western United States and the leading manufacturing centre of the country. The largest industry in the city is the aerospace industry, which makes up about 2000 of the 18 000 firms in Los Angeles. This industry produces airplanes, spacecraft, and parts and equipment related to air and space travel. Another important industry, and one related to the aerospace industry, is the electronics industry. These industries developed in the Los Angeles area because of research programs (often government-funded) conducted by local universities. New technologies developed because of these research programs. **Entrepreneurs** — people willing to risk time and money to set up new businesses — recognized the potential of these new technologies and built industries to produce marketable goods based on them. Many of the jobs in these industries are linked to government defence contracts.

There are other important jobs in this region. More than 60 000 people, for example, are employed in the manufacturing of clothing, producing "California fashions", which are popular throughout the continent. Los Angeles is also home to over 50 record companies. Other locally manufactured products include:

- automobiles
- furniture
- tires
- chemicals
- glassware and pottery
- toys
- travel trailers
- iron and steel

Many industries located in Los Angeles serve the large and relatively well-off western U.S. consumer market. Simple statistics indicate the wealth of this market: for example, one in every five Rolls Royce automobiles in the United States and 11 percent of inground swimming pools are found in the Los Angeles area.

This area has long been famous for its motion-picture industry. Film producers were originally attracted by the sunshine and infrequent rain, important considerations when much of the filming was done outdoors. The industry peaked in size in the 1940s, before the widespread use of television. The industry now produces both motion pictures and television programs.

Figure 4.9 *The Port of Los Angeles*
Why would a port help the economy of a city-centred region?

The economic importance of the port of Los Angeles should not be overlooked. Every day, hundreds of fishing boats return from the Pacific Ocean with catches of fish, including tuna and halibut. This makes Los Angeles a very important fishing port and the centre of one of the largest fish-canning industries in the world. The port also handles ships carrying cargo from around the world. Its value of cargo is second only to that of New York. Its three leading trade partners are Japan, Taiwan, and Korea. In effect, Los Angeles is the United States' window onto the Pacific.

The diverse industrial make-up of the city has created a strong and growing job market. In 1988, there were 7 662 000 jobs in Los Angeles. This figure is expected to increase to about 9 253 000 by the year 2000, an annual increase of about 1.6 percent.

CHECKBACK AND APPLY

12. Identify reasons why each of the following industries finds Los Angeles a suitable place for business:

    a) the aerospace industry

    b) the clothing industry

    c) the movie and television industry

13. The economy of Los Angeles is built on a wide variety of industries, rather than being dependent on just one or two. What advantages does such a broadly based economy have for a city-centred region like L.A.?

14. The port of Los Angeles, besides its fishing and trading functions, also provides docking facilities for small pleasure boats. Identify ways in which these services will benefit the economy of the city.

15. While considering the advantages and disadvantages of Los Angeles, rank the following economic activities in terms of their suitability to the L.A. region on a scale of 1 to 10 (1 is poor: 10 is very good):

    • manufacturing farm machinery
    • tourism
    • automobile manufacturing
    • mining

    Provide reasons for your rankings.

Los Angeles

THE REGION AT A GLANCE

Area	
—city	1204 km²
—metropolitan area	10 653 km²
Population (metro, 1988)	13 770 000
Population density (metro area)	1293 persons/km²
Per capita personal income	
—L.A. (1988)	$15 211
—U.S. average	$13 245
Employment (1988)	7 662 000
Annual rate of employment growth	1.6%
Annual rate of population growth	1.4%
Asian population	1 100 000
Hispanic population	2 800 000
Murder rate per 100 000 people	
—L.A.	88.0
—U.S. average	8.3
Main industries	transportation, high tech, entertainment, tourism, manufacturing
Main exports	machinery, electrical equipment, aircraft, manufactured goods

Tourism in Los Angeles

Tourism is one of the major industries of the Los Angeles region. Built partly on the media image of an attractive, easy-going lifestyle, tourism has become a multi-billion-dollar industry. Consider these facts:

- Close to 105 million people travel through California every year. Of these, 6 million are foreigners, 30 million are Americans from other states, and the rest are Californians touring their own state.
- Over 800 000 Canadians visited California in 1989, most of whom travelled to Los Angeles.
- Tourists spend $5 billion a year in Los Angeles.
- Tourism has created about 85 000 jobs in the region.
- Los Angeles is the largest single region for tourists in California.

Tourists who come to the Los Angeles region can select almost any kind of vacation they want. They can visit beaches, deserts, mountains, forests, farms, or cities. One of the most famous of all tourist destinations is Disneyland, the world's first major theme park. It opened in 1955.

Other attractions in Los Angeles include the Universal Studios tour, the S.S. Queen Mary ocean liner, Mann's Chinese Theatre (where over 160 film stars have left their footprints in the sidewalk), the Walk of Stars in Hollywood, and CBS Television City. A variety of tours parade visitors past the homes of famous movie stars.

For those more interested in the high-brow side of life, Los Angeles has close to 20 orchestras, several opera companies, as well as theatre and dance groups. There are many museums, parks, arts centres, and zoos.

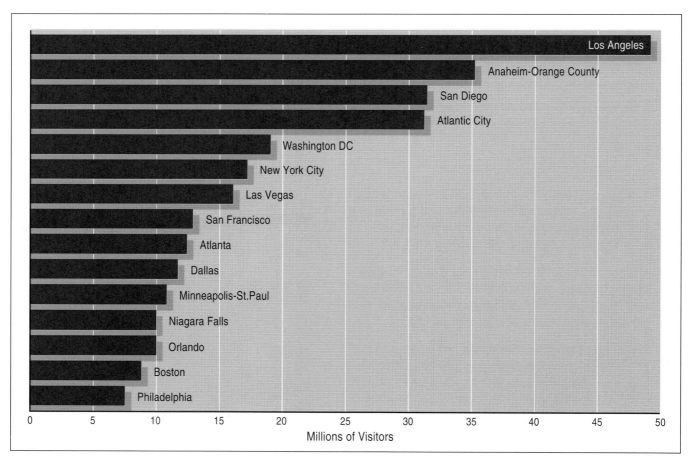

Source: U.S. News and World Report

Figure 4.10 *Number of Visitors to U.S. Cities*
Find these cities on a map of the United States. Are these cities concentrated in specific areas? If so, where?

Los Angeles is famous for its shopping. Rodeo Drive and Beverly Hills are some of the most expensive shopping areas in the world. Hand-sewn custom evening gowns cost more than many Californians earn in one year. By way of contrast, the wholesale district of the city and the flea markets have some of the best values in the United States. Melrose Avenue has over two kilometres of one-of-a-kind shops and intimate bistros; it is one of Los Angeles' trendiest shopping areas.

The beaches in the region are a major attraction, as is surfing, although the water is cool throughout the year. Malibu Beach, Santa Monica, Venice, and Santa Barbara all have outstanding beaches, most of which are accessible to the public.

CHECKBACK AND APPLY

16. What types of attractions would you include in an itinerary for each of the following people who plan to spend a week in the Los Angeles region?

    a) a nineteen-year-old college student

    b) a retired couple who enjoy music but have some difficulty getting around

    c) a young couple with their two children, aged eight and six, with a limited budget

    d) a wealthy, middle-aged couple

17. Design a one-page tourist brochure for the Los Angeles region based on the theme that there is something there for everyone.

18. Tourism is an industry that is sensitive to changes in the economy and in other areas of society. List and explain factors in Los Angeles, in the rest of the USA, and in the world that could affect — positively or negatively — the tourist industry in the city. Explain your choices.

TOURIST GUIDE

- The climate of Los Angeles is mild the year round, with a rainy season between November and March. The heaviest precipitation occurs in January.
- Almost all banks in California belong to a network of automatic teller machines (ATMs), so cash can be obtained 24 hours a day with a bank card.
- You can usually obtain tickets to be part of the audience of television shows by simply phoning the network studio.
- In Los Angeles, having a vehicle is a necessity. The size and character of the city do not encourage walking or the use of public transit. Shop around to get the best car rental rates.
- Bus tours are a good way to get oriented to Los Angeles and to see a number of sights in a short time.
- More Mexican Americans live in Los Angeles than in any other city in the U.S. They help to give the city a highly cosmopolitan flavour.

Famous Tourist Destinations

Beverly Hills

Beverly Hills has a world-wide reputation for its movie stars, fabulous shopping, and luxurious hotels. It is a place where only the rich can afford to live, but others can visit and enjoy its charm.

For the wealthy, Beverly Hills is a shopper's paradise: shopping is seen as an art. The finest department stores — Saks Fifth Avenue, Neiman-Marcus, I. Magnin — are found here, along with exclusive specialty shops. This is an area where you can buy such items as $200 socks and where some stores take customers by appointment only.

The hotels in this area are world-class and are often frequented by visiting dignitaries and celebrities. If cost is not a problem, you could stay at the elegant Beverly Wilshire or the secluded Hotel Bel-Air. There are less opulent accommodations around, but this is not a place for travellers on a budget.

Beverly Hills has been a city since the early part of the century. Its reputation for wealth and luxury has allowed it to retain an identity separate from the larger Los Angeles. Its perfectly manicured lawns and palatial homes set it apart from the work-a-day life of the city.

CHECKBACK AND APPLY

19. In what ways will the wealth of the people of Beverly Hills help the city keep a separate identity from Los Angeles?

20. In this area, hawkers on the street try to sell visitors tour maps of movie stars' homes.

    a) Why do you suppose people buy tour maps of stars' homes?

    b) Do you consider this an invasion of the privacy of movie stars? Why or why not?

    c) If you operated a tour service, how might you conduct tours without invading the privacy of residents?

Figure 4.11 Beverly Hills
Beverly Hills, more than any other place, symbolizes the glamour and excitement of wealth. What does Beverly Hills conjure up in your mind?

The Changing Population of Los Angeles

Since 1970, the nature of the population of Los Angeles has changed drastically. Waves of immigrants — both legal and illegal — have entered the region, as well as refugees, most notably from southeast Asia. The attraction of the good life in Los Angeles, the climate, and the opportunity to work hard and succeed are the reasons for this spectacular population growth.

California attracts close to 30 percent of all immigrants to the United States, and most of that 30 percent comes to the Los Angeles region. Some of these immigrants come, as in the past, from European countries. Many, however, come from Latin America and the Far East. As the region grows in population, the composition of its population is taking on a more Latin flavour.

Figure 4.12 *Growth Rates of Minority Populations in Los Angeles*

Minority Group	Percentage of L.A. Population	Rate of Growth
Hispanics	31.5	32.3
Blacks	12.7	12.1
Asian/Other	9.5	51.2

Source: U.S. Bureau of the Census

In the Los Angeles region, there are close to a quarter million refugees. Most of these people arrived in California as a result of the American involvement in the Vietnam War and the conflicts in other neighbouring countries during the 1970s and 1980s. Figure 4.13 reflects this.

The numbers given in Figure 4.13 do not include the illegal immigrants to the Los Angeles region. These are immigrants who arrive without meeting the U.S. government regulations for entry into the country. The actual number of such people who enter the region is not known, but approximately 500 000 illegals are caught every year near the border between Mexico and the United States.

Figure 4.13 *Major Countries of Origin of Legal Immigrants to California, 1988*

Mexico	53 622
Philippines	25 012
China	11 273
Vietnam	11 096
Korea	9 748
Iran	8 160
El Salvador	6 829
India	4 576
Taiwan	4 300
Hong Kong	3 750
Laos	3 649
Kampuchea	3 289
TOTAL IMMIGRATION TO CALIFORNIA	188 696

Source: U.S. Bureau of the Census

CHECKBACK AND APPLY

21. If current immigration trends to Los Angeles continue, what might the city's population be like twenty-five years from now?

22. a) Some critics have claimed that the United States has lost control of its borders, especially the U.S.-Mexican one. What do you think is meant by that claim?

    b) What are the implications of this if it is true?

Challenges Facing Los Angeles

Water supplies have not kept pace with the rapid growth of the population. As a result, the Los Angeles region is suffering from a severe water shortage. The climate is semi-desert, so there are few rivers flowing through the Los Angeles basin. To cope with the situation, the city has piped in water from as far away as the Colorado River. But even this water has not been sufficient to meet the needs of the city and the irrigated farming in the nearby mountain valleys. Solutions to the problem are expensive. So far the options appear to be importing water by ship, damming rivers and rerouting water from northern California, increased conservation, reducing the amount of irrigated farming, and **desalination** — removal of salt — of sea water. A reliable, adequate supply of water must be found to satisfy the people and businesses of the Los Angeles area if the economy is to continue to grow.

Another difficulty for the region is seismic activity — earthquakes. Several fault lines run through Los Angeles, and when movements occur along these lines, earthquakes take place. In such a built-up area, earthquakes could cause millions, if not billions, of dollars' worth of damage to freeways, buildings, and major facilities. A great deal of effort has been put into building new structures to withstand earth tremors, as well as strengthening existing buildings. Nevertheless, earthquakes will continue to be a worry for the people of the Los Angeles region.

A major problem for the city is racial and ethnic conflict. As you saw in Figures 4.12 and 4.13, there are large communities of ethnic and racial groups within Los Angeles. Unfortunately, many of these people live in run-down and overcrowded parts of the city. A large proportion of Blacks live in south-central Los Angeles, including the district of Watts, while the **barrios** — shacks — of east Los Angeles are home to Mexican Americans. Blacks and Latinos are often prevented from moving out of

Figure 4.14 *A Barrio in Los Angeles*
List five ways in which living in a barrio could hinder you from getting regular employment.

these areas by the prejudice and bigotry of some people in the surrounding communities. The local, state, and federal governments have attempted to improve conditions by setting up legal aid programs for members of minority groups who have been discriminated against, training courses to give workers better job skills, and language schools to help newcomers improve their English. However, progress has been slow because poor, unskilled immigrants continue to pour into the already overcrowded districts of the city.

The frustrations of large-city life take their toll on people in a city. Crime rates are a dramatic indication of this. Los Angeles has the unenviable distinction of having the fourth highest rate of violent crime in the country, and the crime rate is rising quickly. There has been an alarming increase in the number of youth gangs whose violent activities are related to the distribution and use of illegal drugs. This is not a problem that will be overcome quickly or easily.

CHECKBACK AND APPLY

23. What could individual homeowners do to try to conserve water, a vital but limited resource? List at least 10 suggestions.

24. a) Why do some racial and ethnic groups concentrate in certain parts of cities?

b) List disadvantages for cities when racial and ethnic groups form distinct neighbourhoods.

c) Identify three actions that a city government could take to reduce discrimination against visible minorities in its jurisdiction. Give reasons for your suggestions.

25. a) What effect might a high crime rate have on tourism?

b) How might this effect be minimized?

26. Which of the problems mentioned in this section of the chapter do you feel is the most serious threat to tourism in Los Angeles? Explain your reasoning.

The Future of Los Angeles

Los Angeles is the centre of a region that has glamour, wealth, a beautiful setting, and an attractive year-round climate. But these are not the only factors that contribute to Los Angeles' growth. This city is also a major centre for new ideas and for new businesses. Every year, thousands of entrepreneurs establish companies here, manufacturing new products or providing new services. With a market in the Los Angeles basin of over 13 million people, there are plenty of opportunities for businesses to prosper. Past entrepreneurial activities include the opening of the world's first drive-in restaurants and leading the entertainment world with high-tech movies and modern television situation comedies.

Los Angeles is very much a region where new ideas are readily accepted. People are willing to experiment with products that might not gain an acceptance anywhere else in the United States. Trends that have begun in Los Angeles have spread quickly. It is this willingness to be innovative that will successfully carry the Los Angeles region into the future.

CHECKBACK AND APPLY

27. a) What characteristics would a person need in order to be a successful entrepreneur?

b) What benefits would the work of entrepreneurs bring to the Los Angeles region?

Chapter Review

- Los Angeles influences a much wider area than it occupies. For this reason, it may be considered a city-centred region.

- Growth in the region has created a huge urban area composed of a large number of separate cities.

- Freeways are a fact of life in Los Angeles as people move throughout the region to get to and from their jobs.

- Major economic activities in the region include tourism, entertainment, and the aerospace industry.

- The region is growing quickly; the influx of immigrants is changing the characteristics of the population.

- Some of the major challenges facing Los Angeles include pollution, crime, water shortages, and earthquakes.

Vocabulary Review

annexation	freeways
barrios	gridlock
city-centred region	infrastructure
desalination	smog
entrepreneurs	

Thinking About . . .

LOS ANGELES AS A CITY-CENTRED REGION

1. Compare life in the Los Angeles region to life in the area in which you live. What are some of the key differences and similarities? What accounts for these differences and similarities?

2. Using the following headings, briefly summarize the characteristics of the Los Angeles region:
   - Urban Structure
   - Economic Activities
   - Opportunities for Tourism
   - Problems

3. Write a one-page "Brief Guide for Tourists to Los Angeles". What places would you recommend that tourists visit? Give your reasons.

4. Figure 4.15 shows the projected per capita incomes for residents of Los Angeles and for the average U.S. citizen.

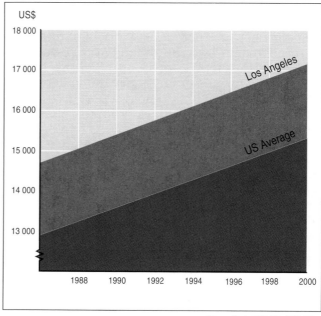

Source: Book of Vital World Statistics

Figure 4.15 *Projected per Capita Personal Income, 1988-2000*

   a) Describe the patterns that the graph illustrates.

   b) How do you suppose the data for this type of graph was generated? What might be some of the problems with a graph of this nature?

   c) If the graph is accurate, what are the implications for Los Angeles? for the United States?

   d) In what ways will per capita personal income levels influence tourism in Los Angeles?

5. Suppose you were the governor of California and you wanted to try to improve the traffic situation in the Los Angeles region. Prepare a three-point plan for tackling this problem. Include in your plan actions that will lead to improvements:
   - within the city of Los Angeles;
   - within the Los Angeles basin;
   - within southern California.

   Make realistic plans, given the technology that will likely be available in the foreseeable future and the costs of improvements. Write a one-page summary that describes your plan.

Atlas Activities

1. a) Refer to pages A26, A27, and A32. Los Angeles has a climate very similar to that of San Diego. Describe the yearly pattern of precipitation and temperature.

b) What difficulties does this type of climate present for farmers?

c) Compare the temperature and precipitation patterns for San Diego with those of the other cities whose climographs appear on page A26. Describe the major differences that you discern.

d) Refer to the world map on page A12. What other places in the world have a climate similar to that of California?

2. Carefully examine the coastline of the Los Angeles area. Where would port facilities be located? Explain your reasoning.

3. This question gives you an opportunity to compare Los Angeles and Hong Kong, both of which are city-centred regions, though in very different areas of the world. Turn to pages A21 and A32. What similarities can you see between the two cities? Think especially in terms of their geographic locations as well as their proximity to other countries.

 ## Further Explorations

1. Select three major tourist attractions in the Los Angeles region and describe:

a) the facilities that they offer tourists;

b) their locations;

c) their backgrounds (in brief);

d) factors that affect their business;

e) types of jobs that are found there.

2. Select any two challenges that face Los Angeles. Find out:

a) the extent of each problem;

b) how each one affects tourism;

c) how the problems could be reduced or eliminated;

d) the future outlook if the problems cannot be controlled.

3. Compare the Los Angeles city-centred region to another type of region. Some headings that you might use are:

- Type of Region
- Size of Region
- Characteristics
- Advantages for Tourism
- Problems

A comparison chart would be a good way to organize your information.

4. Compare the Los Angeles city-centred region to another city-centred region. Choose a city that you are familiar with and/or that you have data on. Compare the regions using three or four different points of comparison.

5. Find out what is involved in becoming a city planner. What kind of background is required? What are some of the issues that city planners have to deal with?

5

The Caribbean
A CULTURAL REGION

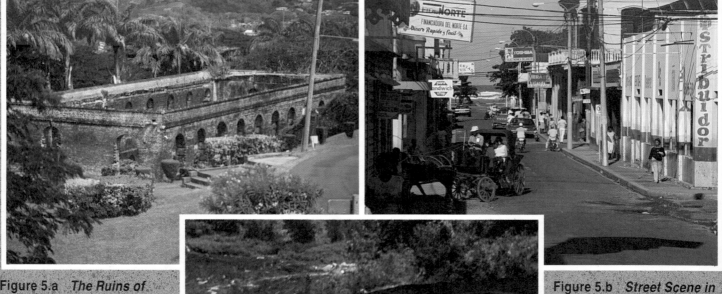

Figure 5.a   *The Ruins of Fort King George, an ex-British Fort in Tobago*

Figure 5.b   *Street Scene in Puerto Plata*

Figure 5.c   *A Bauxite Mine in Jamaica*

Figure 5.d *A Beach in Barbados*

Figure 5.e *Dancers in Traditional Dress in Puerto Plata*

STARTING OUT

1. How could the development of a bauxite mine affect the traditional culture of Jamaica?
2. What can you tell about the physical environment (vegetation, landforms, climate) of the Caribbean region from the photos? How might the physical environment influence the culture of the people?
3. Name three types of economic activity suggested by the photos. How might they be related to tourism?
4. Give two advantages and two disadvantages that the Caribbean's location has for tourism and for manufacturing.

LEARNING DESTINATIONS

By the end of this chapter, you should be able to:
- explain what makes the Caribbean a cultural region;
- appreciate how the area's history has contributed to its cultural traditions;
- suggest how the physical environment has influenced the region's culture;
- understand the economic advantages and limitations of the Caribbean.

Introduction

Visitors arriving at any Caribbean island airport for the first time are almost always amazed by what they see. They find themselves in a place totally different from any they have known. Caribbean islands are unique. A ride from an airport to a hotel vividly reveals this uniqueness, as one Canadian visitor discovered:

> " *The taxi driver chatted to us in rich melodious tones as he drove along a rough road. During that half-hour drive, we saw sights that we will always remember: beautiful houses surrounded by lovely gardens; small shacks with corrugated roofs and open shutters; chickens scratching at the ground; goats tethered on the rough grass. We passed several small shops, a group of people playing cricket, an old man braiding palm leaves into mats, and a bus packed with workers returning to their homes. Near the hotel was a small market, with people selling everything from T-shirts to delicious, spicy foods and fresh fish. We saw several women in brightly coloured dresses walking home with their purchases wrapped in cloths, balanced on their heads.* "

Other tourists have been delighted by similar scenes.

The Caribbean region is distinct because of the backgrounds of the people who live there. The earliest inhabitants were Indians. In fact, the name "Caribbean" comes from the Caribs, one of the earliest **Amerindian** (American Indian) groups to come to the region from South America. In the late fifteenth century, colonists from England, France, the Netherlands, and Spain began to arrive. These colonists established sugar plantations on many islands. People from different parts of Africa were captured, taken as slaves to the Caribbean area, where they were forced to work on the plantations. The descendents of these slaves form the majority of today's Caribbean population.

All of the groups brought with them their own beliefs, values, religions, strong sense of identity, foods, traditions, languages, communication styles, perceptions, and music. Over the years, some of these cultural characteristics changed, perhaps because of the new physical and social context in which the people found themselves. Sometimes traditions of different cultural groups became integrated with those of others. Out of this has developed a special **cultural region** that is a distinctive blend of all the many cultures and influences in the area.

CHECKBACK AND APPLY

1. **a)** In groups, brainstorm a list of words that you think describe the Caribbean region using these categories: physical environment, culture, economic activities.

   b) Compare your work with that of others. What does the Caribbean region seem to be most well known for among your classmates? Why do you think this is so?

2. For what reasons might tourists be interested in visiting a cultural region? Explain your answer.

3. Figure 5.1 shows possible outcomes when different cultures meet. The triangle and the square each represents different cultures.

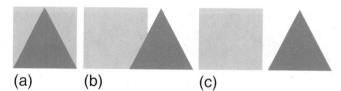

(a)   (b)   (c)

Figure 5.1 *Outcome When Cultures Meet*

   a) Describe what each diagram shows about the meeting of cultures.

   b) The titles "Adoption", "Isolation", and "Integration" could be applied to the three diagrams. Sketch the diagrams in your notebook and give each its appropriate title.

   c) What conditions would favour each of the three cultural outcomes?

   d) Which of the three outcomes do you think best represents the situation in the Caribbean region today? Give reasons for your answer. Refer back to your answer when you have completed the chapter. Has your opinion changed?

The Caribbean as a Cultural Region

The Caribbean region is unique because of the blending of Amerindian, African, and European cultures, with significant infusions of people from India, China, and the Middle East in some areas. The widely different mixes of these cultures has given the area a truly complex set of ethnic origins and languages, as shown in Figure 5.2.

On most islands in the Caribbean region there is a Black majority. The ancestors of these people were brought to the Caribbean as slaves by Europeans. The colonists who had

established sugar plantations on the islands had initially tried to use the native Arawaks and Caribs as slave labour. However, European diseases killed most of these Indians. As a result, the trans-Atlantic slave trade began. Africans from West Africa were captured and transported to the Caribbean region. Between the sixteenth and eighteenth centuries, millions of Africans were taken across the Atlantic. Although uprooted from their homelands and made slaves, the people retained much of their own way of life. Many aspects of the culture of the Caribbean region originated with the African slaves who worked on the sugar plantations.

Figure 5.2 *Ethnic Origins, Languages, and Religions in the Caribbean Region, Selected Countries*

Country	Ethnic and/or Racial Groups	Official Languages	Main Religions
Aruba	80% mixed European and Indian 20% other	Dutch	82% Roman Catholic 8% Protestant
Bahamas	85% Black 15% white	English	29% Baptist 23% Anglican 22% Roman Catholic
Barbados	80% Black 16% mixed	English	70% Anglican 9% Methodist
Cuba	51% mixed 37% white 11% Black	Spanish	42% Roman Catholic 49% none
Dominican Republic	73% mixed 16% white 11% Black	Spanish	95% Roman Catholic
Grenada	mostly Black	English	60% Roman Catholic 20% Anglican
Haiti	95% Black	French	80% Roman Catholic
Jamaica	76% Black 15% mixed	English	mostly Protestant
St. Lucia	93% Black 5% mixed	English	90% Roman Catholic 7% Protestant
Trinidad and Tobago	43% Black 40% East Indian 14% mixed	English	36% Roman Catholic 23% Hindu 13% Protestant

Source: New World Almanac

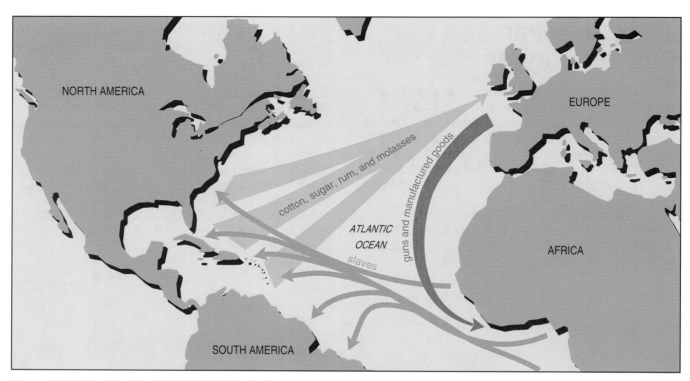

Figure 5.3 *Slave Trade, 16th to 18th Centuries*

In order to communicate with each other, slaves from different parts of Africa, each group with its own language, developed a common language. It was based on the languages of their "masters", but had African grammatical structure, vocabulary, and syntax. This language is **Creole**, a language form that varies from one island to another.

Official languages also vary across the region. In Cuba and the Dominican Republic, the language is Spanish. In Haiti, which once belonged to France, the people speak French. Because the Bahamas, Barbados, Grenada, Jamaica, Trinidad, and Tobago were once part of the British Empire, English is spoken there. Dutch is the official language in Aruba and Curacao. The official language of an island may not, however, be the language of normal use. Local dialects play an important role in day-to-day communication.

When slavery was abolished in the nineteenth century, the freed slaves left the plantations. Plantation owners then brought **indentured labourers** from India, China, and Syria to work the land. An indentured labourer had to sign a contract to work for five years in return for transportation and a meagre wage. Because of unemployment in their home countries, these labourers accepted the harsh conditions and most stayed in the Caribbean area, giving the region an additional cultural dimension.

The blend of all the different cultural backgrounds has given rise to some particularly Caribbean phenomena, such as Calypso music. Calypso music first developed in Trinidad — a result of the combined influences from Asia, Spain, and Venezuela. This type of music is traditionally a ballad made up on the spur of the moment, and concerns current events or man-woman themes. Steel drum bands, another Caribbean specialty, developed in Trinidad during World War II. In search of music, people beat the metal of old oil drums into shapes to create a variety of notes when struck.

Figure 5.4 *A Reggae Singer*
Reggae music is indigenous to Kingston, Jamaica. It has a strong guitar beat and tells of the struggles of everyday life in poverty-stricken areas of urban Jamaica. Reggae music started with the Rastafarians, a quasi-religious group that believes that the only way for Black people to escape their socio-economic plight is to return to Africa.

CHECKBACK AND APPLY

4. **a)** List, in order of arrival, the groups of people who settled the Caribbean islands. Give a reason why each group came to this part of the world.

   b) Give evidence to show that the various groups mixed and blended together to give this region its unique culture.

5. These population figures show the number of Arawak Indians in Hispaniola (now Haiti and the Dominion Republic) following Columbus's arrival there:

1492	300 000
1507	60 000
1509	40 000
1517	14 000
1548	500

   What caused the destruction of this group of people?

6. **a)** Describe how the triangular trans-Atlantic slave trade worked.

   b) Who gained the most from this system of trade? Who lost the most because of it?

7. In what ways is the history of cultural development in the Caribbean region a) similar to and b) different from that in Canada and the United States?

8. How does the historical development of the Caribbean region help to explain why it is economically less developed than North America or European countries?

The Influence of the Environment on Culture

Although the distinctive Caribbean culture is mainly the result of a blend of various cultures, the physical environment of the area also had an influence. The Caribbean islands lie in the tropical zone, a climatic area that is hot and wet with seasonal rainfall. The trade winds are an important climatic control. These moisture-laden winds blow from the Atlantic Ocean to the Caribbean Sea. They give the area high rainfall totals, although windward sides of islands are usually wetter than leeward sides. The trade winds also help to cool the high tropical temperatures and create a very pleasant climate.

Figure 5.5 *Wind Direction Rose for Kingston, Jamaica*
The trade winds blow with amazing constancy, as this wind rose shows. The length of each "arm" indicates the frequency with which the wind blows from that direction.

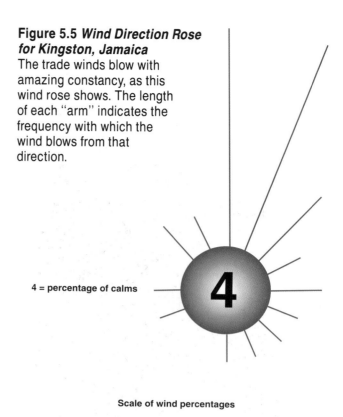

4 = percentage of calms

Scale of wind percentages

| 0 | 10 | 20 | 30 |

Figure 5.6 *Climate Statistics for Four Caribbean Locations*
The first row shows the average monthly temperature in degrees Celsius.
The second row shows the average precipitation in millimetres.

Locations						Months						
	J	F	M	A	M	J	J	A	S	O	N	D
Havana, Cuba (24 m, 23°N)												
Temp. (°C)	22.2	22.2	23.3	24.4	26.1	27.2	27.8	27.8	27.2	26.1	23.9	22.8
Precip. (mm)	71	46	46	58	119	165	122	135	150	173	79	58
Santo Domingo, Dominican Republic (17 m, 19°N)												
Temp. (°C)	23.9	24.2	24.2	25.0	25.8	26.4	26.7	26.9	26.7	26.4	25.6	26.4
Precip. (mm)	61	36	48	99	173	157	163	160	185	152	122	61
San Juan, Puerto Rico (25 m, 18°N)												
Temp. (°C)	23.9	23.9	24.4	25.0	26.0	26.6	26.7	27.1	27.0	26.8	25.9	24.8
Precip. (mm)	104	68	53	99	182	148	152	162	153	133	153	124
Port of Spain, Trinidad (21 m, 11°N)												
Temp. (°C)	24.4	24.4	25.0	25.6	26.1	26.1	25.6	26.1	26.1	26.1	25.6	25.0
Precip. (mm)	58	30	36	33	71	164	198	193	175	142	165	119

In what ways might the climate influence the activities of people in the Caribbean region? How might it influence tourism?

Physically, the islands are the tops of undersea mountains which are connected to the mountain chains of Central America. Many of the islands are volcanic in origin. They tend to have steep slopes, narrow valleys, and deep, fertile soils. The other islands have formed through the buildup of coral **polyps** in the warm Caribbean Sea. Polyps are tiny animals with tentacles, rather like sea anemones. These limestone islands are flat and have shallow, infertile soils. Because of their elevations, volcanic islands receive more rainfall than limestone islands and are better suited to agriculture. It was on these volcanic islands that the European colonists established their sugar cane plantations. The hot, wet climate was ideal for this export crop.

Commercial agriculture is still an important part of the Caribbean economy. Bananas, citrus fruit, coffee, and sugar cane are grown on large plantations and sold to other countries, chiefly the United States, the United Kingdom, and Canada. But subsistence farming — growing food to meet the needs of the family — continues to be practised. Fruits and vegetables are primarily subsistence crops, although small surpluses may be sold at markets or roadside stands.

The continuing dependence on agriculture has shaped the culture of the islands. Many people live in rural settlements, usually hamlets or small villages. A strong sense of community exists among the villagers; the communities provide the people with a social centre. Only in recent years has there been a rural-urban drift of people towards the cities. This is due, in part, to growing rural populations and competition for farmland and the drift of people to large cities in search of jobs.

Figure 5.7 *Traditional Foodstuffs of the Caribbean Islands*

Cacao beans are gently roasted, seasoned with cinnamon and nutmeg, and made into small cakes, which are dried in the sun. These cakes are grated into hot water or milk to make a delicious drink.

Chickens need little space.

Cacao

Goats tolerate poor pasture.

Sugar cane is used to make rum. To make one litre of rum requires sugar cane from a 6 m x 6 m area.

Sugar cane

A few cattle

Fish, especially salted cod fish

Thyme, bay leaves, and parsley

Ackee, mango, sour sop, papaya, citrus fruit, guava, bananas, and pineapple

Arrowroot, Angostura bitters

Coconut

Nutmeg, mace, ginger, cloves, chilies, and annatto

Onions, peppers, tomatoes, plantain, cassava, christophine, yams, sweet potatoes, callalloo, okra, dasheen, pigeon peas, red peas, and breadfruit

Locally raised pigs need no pasture which is scarce.

Grenada is called "the Spice Island." Nearly all the world's allspice comes from Jamaica.

Salted cod fish is a traditional food from the days of slavery.

Immigration from different parts of the world influenced the crops grown in the area. Europeans introduced citrus fruits, such as lemons and oranges. Africans introduced crops such as cassava, yams, corn, and sweet potatoes, which have become staple foods in the diets of Caribbean people. These crops are particularly well suited to the growing conditions in the Caribbean islands.

CHECKBACK AND APPLY

9. Brainstorm a list of advantages of the physical environment of the Caribbean region. Divide your list into economic advantages and lifestyle advantages.

10. In what ways do the prevailing winds and the height of an island contribute to the distribution and quantity of rainfall on that island?

11. a) In what ways might the physical environment of an area affect the culture of the people who live there? Give three specific examples from your own area.

   b) Using the photos in this chapter, and the information given in the preceding section, identify two ways in which the physical environment of the Caribbean region has influenced the culture of the people.

12. a) Using the information in the preceding section of the chapter, suggest how the physical make-up of the Caribbean region is as diverse as its human make-up.

   b) What are some physical characteristics that create a common bond among the islands of the Caribbean region?

JAMAICA, 1988

It was a beautiful sunny morning in September. The sunlight filtered its way through the branches of the papaya, guava, and jackfruit trees that grow in the streets in the quiet town. Moneague, where Kingsley Hurlington lived, is located in the mountains just a short drive southeast from Ochos Rios.

The stillness of the day was abruptly destroyed when a hurricane warning came over the radio. Everyone knew what to do; they had been through this before. The town came to life and the sound of hammering filled the air. People were boarding up windows and tying down loose objects. The local store filled with people stocking up on canned food, candles, lamps, radios, and batteries.

The Hurlington family didn't sleep well that night. With hurricanes you never know what to expect. If this one's course changed, it might not even affect them, and if it was coming their way, they weren't sure when it would hit.

There was no sun visible through the thick, dark clouds that rolled in the next morning. The family huddled round the television, watching the weather reports and waiting. Early in the afternoon, the hurricane hit with high winds and torrential rain. Inside, they heard the thud

Figure 5.8 *Kingsley Hurlington*

Living Through Hurricane Gilbert

of broken branches and whole trees crashing against their walls and roof. The building shook, but survived the battering. The Hurlington home was made of cement blocks and was stronger than many other houses in the area. Soon, groups of soaking wet and frightened neighbours joined them, for their houses were badly damaged, or completely destroyed. The Hurlingtons welcomed over 30 people in all into their home. The power went out and the people lit kerosene lamps and candles, and prepared a meal as best they could.

Suddenly, the wind dropped; the eye of the storm was passing. Everybody burst out of the house to examine the effects of the storm. It was brighter now, and the rain had tapered to a light drizzle. What met their eyes was a scene of utter devastation. People cried when they saw how badly damaged their houses had been and that their crops had been destroyed. The road in front of the house was flooded, and debris, trees, and leaves were everywhere.

The storm returned quickly, and everyone rushed for the shelter of the house. Some slept fitfully as the storm raged on through the night. By morning, the storm had passed, and the task of rebuilding their lives had started. The Hurlington house was an important refuge for many families during the days that followed, until they could rebuild or repair their own houses. In all, about 80 percent of the homes on the island had been damaged. So much of the island had been devastated that it took two months for electricity to be restored to Moneague, and even longer for the roads to be cleared. Within a year, through the determination of the people and much outside assistance, the damage had been partially repaired.

CHECKBACK AND APPLY

13. Briefly describe how the people of Moneague reacted as Hurricane Gilbert went through each of its stages.

14. **a)** What impact would a hurricane such as this have on tourism on the island? Consider both the immediate and the longer-term consequences.

    b) What could be done to minimize the effects of a hurricane on Jamaica's tourist industry?

15. **a)** Suggest five ways in which the Jamaican authorities and/or the citizens might prepare for the next hurricane.

    b) In what ways is it impossible to prepare fully for a hurricane?

Figure 5.9 *Damage Caused by Hurricane Gilbert*
Winds are one of the most destructive forces in a hurricane. Wind speeds of more than 250 km/h were recorded for Hurricane Gilbert.

The Caribbean

THE REGION AT A GLANCE

Because of the physical and human diversity of the region, statistical summaries are given for four representative countries.

Characteristics	Cuba	Haiti	Jamaica	Trinidad and Tobago
Area (km²)	110 860	27 750	10 961	5130
Population (1991)	10 620 100	6 142 150	2 362 000	1 344 600
Population density (persons/km²)	95.8	221.3	215.5	262.1
Urban population (%)	70	29	49	49
GDP* per capita (US$ — 1989)	2000	380	1340	3070
Birth rate (per 1000)	18	45	27	28
Life expectancy (years) —male —female	73 78	52 55	75 78	69 74
Literacy (%)	98	23	82	98
Type of government	communist	republic	parliamentary democracy	parliamentary democracy
Year of independence	1902	1804	1962	1962
Unemployment rate (%)	6	50	19	22
Agriculture (% GDP*)	11	32	9	3
Tourism (% GDP*)	0.5	3	28	4
Main industries	textiles wood products cement	sugar refining textiles cement	rum molasses cement paper	oil products rum cement tourism

*Gross Domestic Product

Tourism in the Caribbean Region

The Caribbean's unique and interesting culture helps to attract tourists. The social and cultural characteristics of the people can be seen as resources on which the industry is built. Tourists enjoy experiencing local customs, traditions, language forms, hospitality, dances, ways of dress, and cuisine. Of course, the physical beauty and pleasant climate cannot be overlooked. This region has the "sea, sand, and sun" that many vacationers want.

Tourism is a very important part of the economy of the Caribbean region. Most tourists come from North America and Europe, although some islands see a substantial intra-Caribbean flow of tourists. Figure 5.10 shows the destinations that Canadian tourists choose. Tourists' choice of destination seems to be influenced by the distance between their homes and their destinations, cultural ties (such as language), historical links, costs, and recent advertising campaigns.

Tourists create jobs for islanders. In the Bahamas, for example, 25 percent of the labour force is employed in the tourist industry, and 70 percent of the GDP comes from tourism. Workers are involved in a great variety of jobs, from transportation and accommodations, to recreation, retailing, and restaurants. Indirect employment related to tourism includes the construction of facilities, the importing of food and other goods purchased by visitors, and landscaping. It has been estimated that a single Club Med project in the Caribbean creates jobs for about 600 people. On some islands, the majority of jobs are related directly or indirectly to tourism. The map in Figure 5.11 identifies the importance of tourism to Caribbean countries.

Figure 5.10 *Destinations of Canadians for Selected Islands*

Countries	1980	1985	1988
Bahamas	125 000	100 000	92 000
Barbados	110 000	87 000	57 000
Cuba	33 000	60 000	67 000
Dominican Republic	na	na	167 000
Jamaica	70 000	77 000	98 000

Source: Statistics Canada

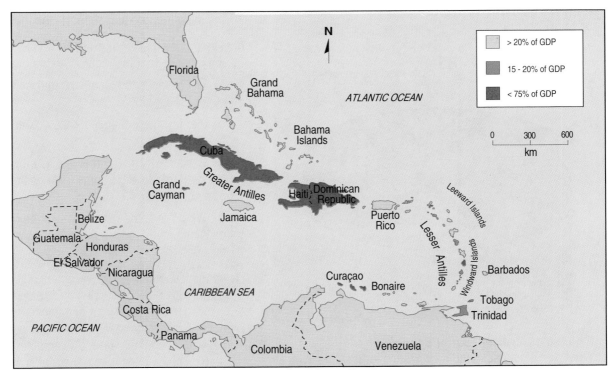

Figure 5.11 *The Importance of Tourism in the Caribbean Region*
What might cause the importance of tourism to vary from one island to another?

Islands benefit from tourism in other ways too. Government revenues are generated through airplane and ship landing fees, departure taxes, hotel bed taxes, and customs duties on imported goods. Governments, however, use much of this revenue to improve the **infrastructure** that supports tourism. The infrastructure is made up of all the facilities and services that tourists require. Governments in most Caribbean islands must spend money on:

- improving airfield runways and installing lighting and navigational equipment;
- expanding terminal buildings and passenger processing facilities;
- building customs and immigration accommodations;
- expanding and improving public utilities, such as electrical and freshwater supplies.

While these improvements are expensive, they are necessary investments for islands that want to attract tourists and tourist spending.

The large capital investment in improving tourist facilities has limited the involvement of local people in developing tourism. It is often foreign investors who put up the capital to build hotels and resorts. Much of the profit from these operations returns to the investors; this is a form of "leakage" of tourism revenue from the Caribbean region. The major benefits to the region's economy come through the employment of local people.

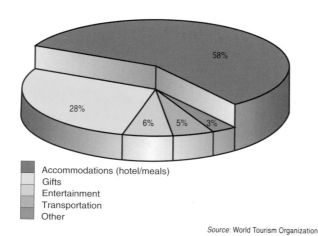

Accommodations (hotel/meals)
Gifts
Entertainment
Transportation
Other

Source: World Tourism Organization

Figure 5.12 *Typical Tourist Spending*
What part of tourist spending plays the largest role in stimulating the local economy?

CHECKBACK AND APPLY

16. Tourism is an important industry in all of the Caribbean islands.

    a) What do you think might attract North American tourists to the region?

    b) What two aspects of tourism might local residents like? What two aspects might they dislike?

17. Describe four factors that might influence people's decisions about vacation destinations.

18. Figure 5.10 shows that the number of tourists visiting Caribbean islands has changed in recent years.

    a) Suggest three reasons why the number of tourists to a place might increase over a few years.

    b) What are three reasons to explain why the number of visitors to an island might decrease?

    c) In chart format, identify the effects on an island's economy and culture when tourist totals increase and when they decrease.

    d) Suppose an island's tourist industry was having difficulty attracting tourists. What three actions might the government take to try to attract more visitors?

19. Explain why the government of a Caribbean island should use tax dollars to cover the costs of providing the infrastructure that supports the tourist industry.

20. The money that tourists spend while in a country is the revenue that the industry earns.

    a) List seven to ten local jobs that will be stimulated by tourist spending.

    b) Identify five leakages for the country's tourist revenue.

    c) Explain why governments of Caribbean countries are prepared to allow some leakages of revenues to other countries.

    d) What are some ways in which the government of an island might reduce the leakages you identified in part b)?

21. Imagine yourself as a government official on a Caribbean island. Describe three safeguards you would use to protect your economy from exploitation by foreign investors. Remember that you are dependent on these investors and that you want to reap as much benefit from the tourist industry as possible, in terms of both money and employment.

Famous Tourist Destinations

Exploring a Coral Reef

The warm waters of the Caribbean encourage the growth of coral reefs. Coral consists of many varieties of polyps. The polyps' tentacles absorb oxygen and nutrients from the sea water. Each polyp secretes a protective shell around its soft body, and it is these shells that build the reef. One polyp grows and builds on top of its parent, and so the reef grows. Between the branches and clumps of coral live an assortment of brightly coloured fish and other creatures, some of which have poisonous spines.

To see the beauty of a reef, you need to look below the surface of the water. You can do this in a glass-bottomed boat or by using swim goggles, but perhaps the best method is to go scuba diving. The word "scuba" consists of the initials of "self-contained underwater breathing apparatus". If you rent the equipment and learn how to use it safely, you will be able to explore the reef at close quarters.

Besides their importance as a tourist attraction, reefs protect the beaches and coastline from excessive wave erosion. They are often damaged by hurricanes, but a more recent threat has come from human activities. Many people break off pieces of the coral, either to keep for themselves or to sell. In addition, where large areas of island slopes have been cleared of forest, as in Haiti, the

Figure 5.13 *Diving to View the Coral Reef*
Why might tourists find diving in the Caribbean region an intriguing experience?

sediment that washes into the rivers and then into the ocean literally chokes the coral, so that it dies. Coral needs clean, oxygen- and nutrient-rich salt water to flourish. Some tourist hotels discharge sewage directly into the sea, further damaging the reef. Thus, eventually, the very beaches that attract tourists may be destroyed by the indirect effects of the tourist industry.

Another imbalance, partly attributable to tourism, is due to the harvesting of conches (large shellfish) for food and for their attractive shells. The conch is a natural predator of a particular sea urchin that kills coral. With fewer conches to kill the sea urchins, coral reefs in some areas are being destroyed, and the reefs no longer continue to grow.

CHECKBACK AND APPLY

22. What types of jobs would be stimulated by tourist enjoyment of coral reefs?

23. a) List the forces that are destroying coral reefs.

    b) For each of these forces, identify one course of action that could be taken to protect the reefs.

TOURIST GUIDE

- Visa and passports are not required to enter some Caribbean countries; however, it is strongly recommended that you carry a passport while in a foreign country as this is your proof of citizenship.
- Fresh water is in short supply on some islands during some times of the year. Expect to reduce your personal consumption.
- Fresh shellfish and fish are readily available throughout the region at very reasonable prices. Many excellent local dishes use fish as the main ingredient.
- There is strong competition among hotels and other facilities for tourists, so it is wise to shop for the best values and compare prices.
- If you wish a quiet holiday, smaller, more remote islands will be better choices than the more popular destinations. Antigua, Montserrat, Dominica, and St. Kitts and Nevis are good places for quiet comfort.
- Purchases of local crafts and art products directly benefit an island's economy. Enquire about the origins of gifts you are purchasing and consider how you can most effectively boost the economy of the region.

The Future of the Caribbean Region

Several problems in the Caribbean region have defied solution and will likely continue to affect the prosperity of the area.

Like many developing parts of the world, the Caribbean region is dependent on agriculture, particularly the production of sugar cane. Fluctuations in the world market because of oversupply or falling demand mean that producers earn less for their crops. This can hurt businesses and other activities that rely on farm income. Tourism is one way of diversifying the economies of Caribbean countries to protect against downturns in agricultural prices.

While tourism does provide jobs, they tend to be seasonal, and there are periods of times when work is scarce. These periods are typically in May and June and September and October, when weather conditions in North America and Europe are somewhat favourable and vacation periods have not yet begun. In Antigua, for example, almost one-third of workers lose their jobs during the off-time for tourists.

Even during peak seasons, Caribbean countries may not realize the full benefits of the tourist trade. Luxury hotels tend to buy their supplies from overseas suppliers, usually in frozen or packaged forms. Often the better paid, more skilled workers in these facilities are non-residents who stay for limited periods of time and then return to their homes. Decisions these people make are sometimes not beneficial to the economic or cultural life of the community. In some instances, hotels built to serve tourists have restricted the access of local people to beaches and other facilities.

Some people from the region think the influx of tourists is harmful to the Caribbean way of life and culture. Rural people are attracted to the tourist jobs in the urban centres, and the rural communities suffer from the loss of these people, who are usually young and ambitious. Traditional forms of music, dance, and art are commercialized in order to satisfy demands of

Figure 5.14 *Hotel in Jamaica*
What are the advantages and disadvantages of working at a hotel such as this?

tourists, often debasing the traditions from which they sprang. The meeting of cultures does not necessarily result in improvements to the local one.

On the other hand, tourism has helped to raise living standards in Caribbean countries. The revenue from tourism is a form of "invisible" exports: it is as if goods had been sold to people from other countries. The money that is earned stimulates the local economy and encourages the development of facilities such as schools, hospitals, and water treatment plants. The income from tourism also allows areas of natural beauty to be preserved as national parks or game reserves. In addition, the people of the area acquire skills in such occupations as hotel management and air transportation.

In recent decades, people from the Caribbean region who are dissatisfied with local prospects have been coming to Canada and the United States in search of better opportunities. In part, this movement is a result of the increased awareness of these countries through tourist contacts. While the young, skilled, enterprising people who leave the region benefit the North American countries, they leave the Caribbean region less able to meet the demands of the future.

It is clear that a number of economic and social problems will continue to challenge the Caribbean region for the years to come. It is equally clear that tourism will be important to the economies of this area in the short- and long-term, given the absence of viable alternative sources of economic activity. It is to be hoped, however, that the government and people of the area can accommodate cultural adjustment in a way that encourages the tourist industry while promoting jobs and economic stability.

CHECKBACK AND APPLY

24. Identify and describe four problems faced by people in the Caribbean region.

25. Suppose you were faced with the decision to leave your rural community to go to a city to work in a tourist hotel.

    a) List "push" factors that would encourage or force you to leave your home community to go to the city.

    b) Identify "pull" factors that would attract you to the city and the job in the hotel.

    c) What factors would encourage you to stay in your home community?

26. Is tourism good for the Caribbean region? To help you answer this question, brainstorm lists of advantages and disadvantages of the tourist industry for the region. Arrive at a conclusion and present your answer to the question in a paragraph.

Chapter Review

- The Caribbean may be regarded as a cultural region because of its unique blend of cultures and races.
- This blend of cultures is the result of the settlement by Native peoples, colonization by Europeans, use of slaves from Africa, and the immigration of labourers from other parts of the world.
- The rich mixture of cultural groups has resulted in the development of special language forms, traditions, attitudes, foods, music, and lifestyles.
- The colonial influence of Europeans is evident in the language, education, and government of the countries of the region.
- The region's unique culture, pleasant climate, and attractive physical environment have made it a popular tourist destination.
- Tourism plays an important role in the economy of the region.

Vocabulary Review

Amerindian
Creole
cultural region

indentured labourer
infrastructure
polyps

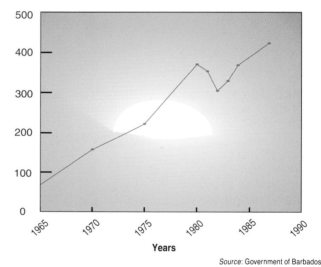

Thinking About ...

THE CARIBBEAN AS A CULTURAL REGION

1. **a)** Define the term "cultural region".

   b) Explain why the Caribbean can be considered a cultural region.

   c) List three other cultural regions you think exist.

2. Identify the cultural groups that came to make up the Caribbean culture. Which groups do you think are very influential? Explain your answer.

3. **a)** Describe three specific ways in which the physical environment has influenced the culture of the region.

   b) Why are the flat limestone islands of the Caribbean usually very dry with a shortage of drinking water?

4. Examine the data in Figure 5.15 about the number of visitors to Barbados.

   a) Describe the pattern that the graph shows.

   b) What might account for the sharp downturn in the number of visitors in the early part of the 1980s?

   c) Identify five specific ways in which this slowdown in tourism would have affected the economy and the people of Barbados.

Thousands of Visitors

Source: Government of Barbados

Figure 5.15 Tourist Visits to Barbados, 1965-1987

5. Plan an advertising campaign to attract Canadian and American tourists to the Caribbean region. Include in your plan:

   a) a clear goal statement

   b) a list of resources that could help you achieve your goal

   c) problems that you will have to overcome in order to be successful

   d) steps that you will take to put your plan into action

   e) a list of ways in which you can evaluate your plan

LOOKING BACK AT THE CARIBBEAN REGION

Atlas Activities

1. Compare the Caribbean and Indonesian islands shown on page A10. Consider the geographic location and the physical, geological, and other characteristics. Use the world thematic maps on pages A4-A14 for additional information.

2. Refer to pages A28-A29. Locate the Panama Canal on both the main and the insert map.

   a) If you were travelling from the Atlantic to the Pacific Ocean, in which compass direction would you travel along the Panama Canal?

   b) What advantages do you consider this route has over possible others through Central America for the construction of a canal?

   c) Why does this canal have locks, while the Suez Canal in Egypt has none? (See page A15 for a map that includes the Suez Canal.)

3. Imagine yourself on a journey from the Atlantic to the Pacific aboard a ship travelling at an average speed of 15 km/h through the Panama Canal. Write a one-page journal entry recording your experiences and observations throughout the journey. The maps of climate and vegetation on pages A12 and A13 should help.

Further Explorations

1. **a)** Identify an island (such as Trinidad, Jamaica, Grenada) in the Caribbean region that has a rich blend of different cultures. How is such richness portrayed in the island's language forms, religion, architecture, foods, music, and sports?

   b) The Caribbean region is rich in musical forms and content. For example, calypso is a social commentary on everyday life; rap and reggae disseminate political messages; and the steel band is versatile enough to play anything from Bach to Beethoven. Bring in samples of such recordings and discuss their characteristics with your classmates.

   c) Research an activity that is typical of the life of people living in the Caribbean. You might be interested in a particular craft, sport, or game. Give as practical a demonstration as is possible. Explain the origins of the activity and its importance to island life.

2. With a partner, plan a two-week visit to the Caribbean region. Use travel brochures and other sources of information to get the details that you need. Include at least three islands in your itinerary, plus a detailed map. Give reference to such practical matters as costs, travel dates, and documents needed. Make suggestions for clothing and other equipment that you will need.

3. Investigate the cruise industry as a potential employer. Find out and report on:

   a) the kinds of background and/or training that are needed

   b) what would be involved in several of the occupations

   c) the salaries and benefits you might expect

   d) your likes and/or dislikes about the types of jobs

   e) your conclusions as to whether or not this is worth considering as a future career

4. If possible, interview a person from the Caribbean region about his or her experiences as an immigrant. Include the following topics in your written report:

   a) the personal background of your subject, including when he or she came to Canada

   b) the reasons the person had for leaving his or her homeland

   c) the features that attracted this person to his or her new home

   d) the person's feelings upon arrival in Canada

   e) the desires of the person to return to the Caribbean

   In your report, suggest ways in which you as an individual could help immigrants settle into an environment that may be unfamiliar and puzzling.

6

Peru
A HISTORIC CULTURAL REGION

Figure 6.a   *Central Plaza in Arequipa*

Figure 6.b   *Indian Village on a Floating Island of Reeds in Lake Titicaca*

Figure 6.c   *A Market in the Highlands of Peru*

Figure 6.d *A Roadside Vendor Near Pisaq in Central Peru*

Figure 6.e *Llamas, Near Ambo in Central Peru*

STARTING OUT

1. Describe the location of Peru in relation to:
   - the rest of South America;
   - your home.
2. Describe the variety of Peruvian scenes shown in the photos.
3. What clues do the photos give about Peru's culture?
4. In what ways does the built environment (constructed by people) of Peru differ from the built environment of your community?
5. List five characteristics of Peru that you think tourists would find inviting.

LEARNING DESTINATIONS

By the end of this chapter, you should be able to:
- recognize the unique features of Peru that came about through its historical development;
- identify the cultural characteristics of Peru and relate them to historical circumstances;
- describe features of Peru's cultural landscape that attract tourists;
- recognize some important problems faced by the people of Peru, and identify some possible solutions.

Introduction

Peru is a land of contrasts. A drive of a few short hours will take you from deserts to tropical rainforests, from coastal plains to steep mountain slopes, from crowded cities to remote meadows. These contrasts are some of the first impressions tourists have of the country, but they do not tell the whole story.

Peru has a unique culture. Its roots lie in the Inca empire that flourished in this part of South America from the eleventh to fourteenth centuries, and in the Amerindian groups which preceded the Incas. By 1535, Spanish explorers and colonists had conquered the Incas, but they were never able to extinguish the Native culture, whose roots remain strong. In this chapter, we consider Peru as a **historic cultural region**.

The designation of Peru as a historic cultural region contrasts somewhat to the Caribbean cultural region detailed in Chapter 5. The Caribbean culture was formed quite recently in the world's history, and is a blend of many cultures that came together to form a new one. The culture of Peru, on the other hand, can be traced back many centuries. While the Peruvian culture has been influenced and modified by a number of other groups, it has a continuity that has existed for a very long time. Hence, Peru is a historic cultural region.

Although it is practical to use current political boundaries to discuss historic patterns, it does create a problem. The geographic area occupied by ancient cultural groups changed over time. Often this was because of the changing power and influence the groups had, or because of environmental circumstances, such as drought, that affected where food could be produced. It is difficult to collect data when the geographic area is not clearly defined. For this reason, Peru will be used to define the region even though, in some cases, the conclusions that can be drawn may not apply equally throughout the country.

Figure 6.1 *Woman in Traditional Dress*
Traditional ways of dressing and working are still seen in Peru today.

CHECKBACK AND APPLY

1. **a)** Define the term "historic cultural region".

   b) For what reasons is Peru defined as a historic cultural region?

2. Suppose you were to travel to Peru with no other knowledge of the country other than that contained in the introduction to this chapter. What evidence would you expect to find to show that this country is a) a physical region, and b) a historic cultural region? List seven to ten pieces of evidence for each.

Peru as a Historic Cultural Region

The first people came to the area that is now Peru about 12 000 years ago. Scientists theorize that this movement was simply an extension of the occupation of North America by people who originally came from Asia to Alaska via an ice-age land bridge. Originally hunters and gatherers, the early people of South America learned to grow crops and raise animals. Among their more important products were potatoes, beans, squashes, chilis, and llamas. There is evidence that the people irrigated their fields and developed skills in using metals.

The Chavin Indians were one of the first civilizations in the area. This group reached its greatest influence about 900 B.C. It was followed by a succession of other powerful groups, including the Mochica, Tiahuanaco, and Chimu Indians. About A.D. 1200, a group of tribes called the Incas took control of the valley of Cuzco in southern Peru. Over the next 300 years, they built a great empire that extended for 4000 km along the Pacific coast. The Inca empire included parts of present-day Colombia, Ecuador, Peru, Chile, and Argentina.

The foundations on which the Inca empire was built can still be seen today in one form or another. One foundation was an efficient road network for the rapid movement of goods and armies. There were two major highways that ran north-south, one along the coast and the other through the mountains. Hundreds of smaller, secondary roads joined these two routes. Some of these roads were simple paths notched into the sides of mountains, while others could handle large volumes of traffic. Messages were sent throughout the empire by runners, and freight was carried by animals such as llamas. Despite the huge size of the empire and the volume of commerce that occurred, Incas never used wheels to move any major vehicle. Nor did they have any written words or numbers. Rather, they kept records on strands of knotted yarn.

Figure 6.2 *Terraces Near Machu Picchu*
Terraces of this type were constructed by the Incas to increase the amount of land that could be used for food production. How is a terrace built?

The Incas were able engineers. In addition to building roads, they carved the steep slopes of the Andes into **terraces** (steps built into the sides of mountains) so that farming could take place. They also built a vast network of irrigation canals to supply water to the terraces and to the flat valley floors. Many of these terraces are still used today by Peruvian farmers, and a few of the irrigation systems are still in operation. (See Figure 6.2.)

Another significant achievement of the Incas was the construction of the capital city of Cuzco. All of the buildings in this centre of power were built of squared stone blocks fitted precisely together without mortar. One of the blocks was 10 m high. The blocks were moved using ropes, rollers, and ramps. The Temple of the Sun in Cuzco was the holy centre of the empire for the Incas.

As the Spanish began to explore the west coast of South America in the 1520s, their attention was drawn to the Inca empire with its great wealth in gold and other resources. Francisco Pizarro led the attack on the Incas in 1532. Use of advanced technology, particularly horses and horse-drawn vehicles, helped the Spanish to take over the empire quickly. In addition, the Spanish had an ally — the diseases that they brought from Europe. Smallpox, measles, and malaria, along with other diseases, wiped out millions of Incas and allowed the Spanish to march unopposed through much of modern-day Peru. Internal power struggles among the leaders of the Incas also helped to weaken the empire's defences.

Figure 6.3 *Sacsahuaman Ruins*
The Sacsahuaman Ruins near Cuzco are still impressive in their size and style. In some cases, the stones fit so well together that it is impossible to insert a piece of paper between the blocks.

Once in power, the Spanish began to exploit the riches of the area. Thousands of colonists arrived from Spain to set up farms and run mines. The silver mines at Potosi produced so much wealth that the Spaniards even shod their horses with the precious metal. Indians were forced to work in the mines and on the farms. With this cheap labour, the area quickly became one of Spain's most profitable colonies.

This period of rapid exploration and colonization of the New World by European nations has been characterized as a time of "gold, glory, and god". The Spaniards were motivated by a desire for riches, and by a missionary zeal. The Indians of Peru were forced to become Christians and to take Spanish names. Control of the spiritual lives of the Indians gave the Spanish rulers great influence over the people and helped to keep rebellions from starting.

Peru became independent of Spain through a series of military actions between 1820 and 1826.

CHECKBACK AND APPLY

3. For what length of time has Peru been occupied by Amerindians? by people from other parts of the world?

4. a) What legacy did the Incas leave for future generations? List five specific examples.

b) Which of the points listed do you think is the most important? Why?

5. a) What circumstances allowed a small Spanish force to defeat the Inca empire?

b) What are some techniques that the Spanish used to control the Indians?

c) Suggest some ways in which the rule by Spaniards changed the day-to-day lives of the people of Peru.

d) In what ways would the culture of the Indians have been relatively unaffected by the Spanish conquerors?

The People and Culture of Peru

Even after Peru gained its independence from Spain, people of Spanish ancestry controlled the economic and political life of the country. They owned and operated most businesses, held military posts, filled political offices, and directed religious activities. This small upper class controlled a huge lower class of Indians. The strict separation of the two classes continued until into the twentieth century. Only then did a middle class of managers and merchants emerge, most of whom were white or *mestizos*, people of mixed European and Indian blood. Today, only 15 percent of the population of Peru is white, 45 percent is Indian, and 37 percent is *mestizos*.

People in the few families that make up the upper class in Peruvian society seldom mix socially with other classes, and there are many intermarriages among the families of the rich. Many of the upper class live in Lima, the largest city in Peru, or in other large cities near the coast. The upper class speak Spanish and dress much as people do in other Western countries.

Most of the Indians live in the highland areas of the country and on the coast. The highland Indians live by farming and cultivate land as high as 4500 m above sea level. Many of these people have retained traditional forms of clothing and maintain Inca beliefs and practices under a veneer of Christianity. Indians in the coastal areas are employed on plantations or work in the cities. A smaller group of Indians still lives in the eastern rainforest of Peru, virtually untouched by Western influences. While it is useful to talk of Indians as a single cultural group, they are, in fact, a mosaic of different cultural and linguistic groups. Each group has attempted to maintain its own culture and to separate itself from outside influences, particularly Spanish ones.

While the wealthy people of the country eat a variety of foods, the diet of the poorer people is bland and monotonous. Highland Indians grow potatoes, beans, squash, wheat, and corn. Indians in the rainforest rely on corn and cassava, a starchy root. However, they are able to supplement their diet with fruits, nuts, fish, and small game. Many Indians chew the leaves of the coca plant to relieve hunger pangs.

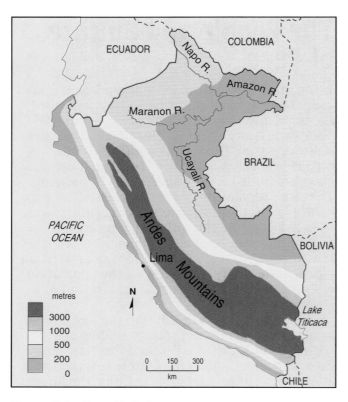

Figure 6.4a *Peru: Relief*

Figure 6.4b *Peru: Precipitation*

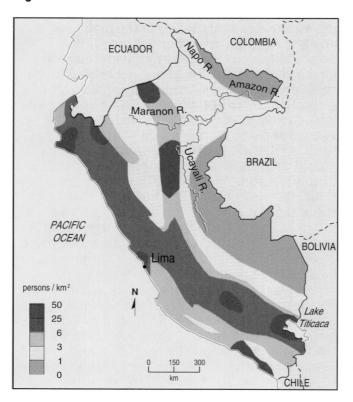

Figure 6.4c *Peru: Population Density*

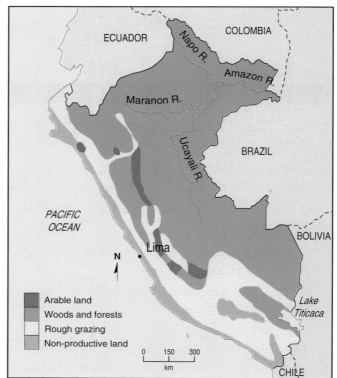

Figure 6.4d *Peru: Land Use*

The coastal area of Peru is arid; it is the fertile valleys of the Andes that produce most of the crops for the nation.

Figure 6.5 *A Squatter Settlement and a City Home*

The official languages of Peru are Spanish and Quechua, an Indian language. About three-quarters of the population can speak Spanish and it is the working language of the country. About 2 million highland Indians speak only Quechua. Scattered groups of Indians in remote areas speak a variety of other tribal languages.

A large proportion of the people of the coastal areas live in poverty. They cannot afford to buy or rent housing, even if it were available. Consequently, many of these Peruvians live in crowded, unsanitary squatter settlements, called *pueblos jovenes* (new towns), on the edges of the large cities. They build homes of old sheet metal, cardboard, and lumber, and travel into the city in search of jobs. Eventually, many of these families can afford to build permanent homes of adobe (sun-dried clay or mud) or concrete blocks. In some cases, governments have supplied the *pueblos jovenes* with running water and sewage systems. In total, more than a million Peruvians live in these squatter communities.

Much of the interior of the country remains poor and isolated from the outside world; life continues there much as it has for untold generations. The imprint of the Incas remains strong in the highlands. Along the coast, however, the Spanish influence predominates. These two cultures stand in sharp contrast to one another and have not grown together.

CHECKBACK AND APPLY

6. Compare the Spanish and Indian populations of Peru using a comparison chart. Some headings you might use in your chart are: social class, location, and occupations. What other headings could you use to compare these groups?

7. a) How have the people of Peru managed to keep their cultural groups separate and distinct?

   b) What problems might develop if groups do not communicate with each other? Is there evidence that such problems exist in Peru? Explain your answer.

8. In what ways does the statement "Today's way of life is often rooted in the past" apply to Peru?

9. Suppose a person of Spanish ancestry and a highland Indian were asked to give their opinions about problems that the country faces and solutions to those problems.

   a) Identify three things that each person would value highly.

   b) In what ways might the two opinions differ?

   c) Which point of view would you tend to support most? Give reasons for your position.

Tourism in Peru

Peru's tourist industry is built on the cultural legacies of the Indians and the Spanish. The Spanish influence is seen most clearly near the coast in the larger cities. Lima has fine examples of colonial architecture, including the Cathedral (which houses the remains of Francisco Pizarro) and the Presidential Palace. As Lima was the centre of religious authority during the time of the Spanish, it has a number of historic churches, the most impressive of which are the San Francisco and San Pedro churches. The port city of Callao is guarded by the fortress Real Felipe, a structure originally built to protect the area from English and Dutch pirates.

Most of the great Indian tourist sites are located inland, in the highland areas of Peru. Cuzco has a combination of pre-Inca and Inca ruins. Unfortunately, many have been built over by the succeeding Spanish and Peruvian populations. One of the best-known structures of the Incas in Cuzco is the Temple of the Sun, a building that has been restored in recent years. Other Indian sites of archaeological importance are located in Sacsahuaman, Ollantaytambo, Machu Picchu, and on islands in Lake Titicaca. These sites play host to numerous scholars, as well as tourists, each year.

The tourist industry has been growing rapidly in the last several decades. In 1970, there were only about 130 000 tourists; by the late 1980s, well over 300 000 tourists a year

Figure 6.6 *Main Cathedral, Lima*
After more than 400 years, many of the Inca structures have been destroyed or built over and examples of Spanish architecture, like this cathedral, predominate. What are some ways in which historical buildings can be protected?

were arriving. This was partly due to improvements in tourist facilities within the country, and partly to increased archaeological activity that has sparked interest in Indian cultures around the world. The government has attempted to capitalize on this interest by setting up government agencies to manage physical and cultural resources, and by giving tax incentives for investment in the tourist sector.

Peru's historic culture is not the only attraction for tourists. The coastal waters of Peru are famous for big-game fishing, and lakes and streams have been stocked with trout. In Lake Titicaca, trout weighing up to 21 kg have been caught (the average weight of a trout is 10 kg). The country also has beaches to enjoy, mountains to climb, and white-water rivers to raft. Tourist authorities are developing tourist resorts, trekking holidays, and mountaineering facilities to take advantage of the physical environment for tourism.

Peru
THE REGION AT A GLANCE

Area	1 285 220 km²
Population (1990)	21 905 600
Population density	17.0 persons/km²
Urban population	67%
GDP* per capita (1989)	US$880
Religion	95% Roman Catholic
Population of largest city	Lima — 5 660 000
Population growth (1983-88)	2.6% per year
(1990-2010)	2.0% per year
Population below poverty line	49%
Years of compulsory schooling	6
Number of elementary students per teacher	35
Tourism	
—# of tourists (1988)	320 000
—earnings	US$300 000 000
Inflation rate (1989)	2775%
Labour force	
—agriculture	35.1%
—industry	16.4%
—services	38.2%
Women in the labour force	26%
Main industries	mining, petroleum, fishing, textiles, clothing
Main exports	fishmeal, cotton, sugar, coffee, copper, iron ore

*Gross Domestic Product

CHECKBACK AND APPLY

10. a) What types of tourist facilities should the government provide for people who wish to visit Inca ruins?

b) In what ways might tourism lead to damage to important ruins?

c) What are five actions that the government could take to lessen the harmful effects of tourism on the historic cultural sites?

11. How do you suppose Indian farmers in the highlands of Peru regard tourism? Explain your answer.

12. The tourist industry in Peru is developing in two quite different areas — the historic cultural one and the leisure and sporting one.

a) Which of these two areas do you think has the greatest potential for tourism in Peru? Why?

b) Is it possible to serve both types of tourists using the same facilities? Why or why not?

TOURIST GUIDE

- The rainy season is from October to April: there is heavy rainfall in the tropical forests.
- Temperatures decrease with elevation, so that average temperatures are about 11°C lower in the Andes Mountains than on the coastal plains.
- Lack of oxygen in high altitudes may bring on headaches and nausea. Visitors to the Andes should rest 12 hours or more the first day in the mountains.
- Immunizations for typhoid, tetanus, measles, mumps, polio, and yellow fever should be kept up to date. Sanitary conditions in some parts of the country are not the same as the standards most North American people are used to.
- It is against the law to take Inca or colonial articles out of the country. Replicas are widely available as souvenirs.
- Three key ingredients of Peruvian cuisine are chili peppers, garlic, and lemon. Most restaurants recognize that visitors are not used to spicy foods and will soften the impact of these ingredients.

Machu Picchu, referred to as the "lost city of the Incas", is one of the most spectacular historic sites in the world. It is perched on a ridge close to 1000 m above the Urubamba River, a branch of the Amazon. Hiram Bingham came upon Machu Picchu in July, 1911. He pulled aside the cover of thick tropical forest to reveal the ruins of the ancient civilizations of the Incas. Machu Picchu had remained unexplored by the outside world since the sixteenth century.

A short distance from Cuzco is the Sacred Valley of the Incas, an area containing the ruins of several ancient complexes, including Machu Picchu. The best way in for tourists is via the Cuzco-Santa Ana railway, which takes you into the heart of the mountains and the tropical forest. The tracks run along the edge of the river, which roars with white water. Close to Machu Picchu the train enters a very narrow canyon, and the forest seems to close in around the train. There is barely room for the railway tracks on the bank of the river.

Figure 6.7 *The Road to Machu Picchu*
The narrow, zigzag road that takes visitors from the river to the site of Machu Picchu was cut into the side of the mountain.

Famous Tourist Destinations

Machu Picchu

The train stops in what seems to be the middle of nowhere; there is nothing to be seen except trees covering the steep side of the mountain. It was this inaccessible location that hid Machu Picchu from the Spanish for over 400 years. A hair-raising, zigzag road up the side of the mountain takes visitors to the tourist hotel, and from there it is a short hike to the ruins themselves. The view is worth every minute of the journey!

From the site, it is almost a thousand metres straight down to the river. On top of this narrow ridge, the Incas built a religious centre out of large rocks cut to fit perfectly together. There is a narrow road which leads to the north and the south along a niche cut into the side of the mountain walls. Four centuries ago, Inca runners followed these tracks to deliver messages to other centres of the empire. Today the trail is used by hikers.

Excavations since 1912 have revealed an exciting glimpse into the lives of the Incas. Most of the buildings, which were less than three stories high, were built in tiers to take advantage of the steep slopes. Terraced fields, warehouses for food, and aqueducts to supply the crops with water are located nearby. All of this suggests a society with a strong central government and a healthy economy. The ceremonial importance of Machu Picchu indicates the importance of religion in the culture.

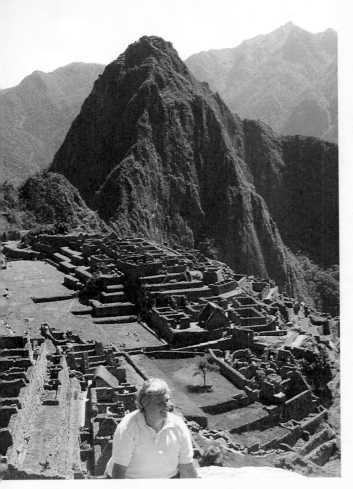

Figure 6.8 *The Ruins at Machu Picchu*
Machu Picchu was not really lost when Hiram Bingham came upon it. The local Indians were well aware of its existence.

CHECKBACK AND APPLY

13. The ruins of Machu Picchu have caught the imagination of the world. They have become well known through travel posters, television programs, and movies. What would you find appealing about this site?

14. What age group or background of tourist would find Machu Picchu most attractive? Give reasons for your answers.

15. a) Based on what you have just read and on your knowledge of tourism, what jobs for local people do you think would be generated by tourists' visits to Machu Picchu?

    b) What might make local people unhappy about the tourist traffic through these ruins?

Famous Tourist Destinations

The Nazca Drawings

To the south of Lima, along the Pacific coast, is an arid area where some extraordinary designs have been discovered. In this area of Peru, it seldom rains and so there is almost no vegetation on the ground to cover the sand. Even so, these designs are not readily visible from the ground. From the air, however, these giant patterns laid out in the sand are clearly visible. Many are geometric in design, while others resemble animal and bird shapes. Some lines are perfectly straight and over a kilometre in length.

The origins of these designs is uncertain, although they have been dated to around the year A.D.500. The purpose of these designs is not known since the builders left no written record. People have speculated that there was some astrological purpose behind their construction.

Figure 6.9 *A Portion of the Nazca Designs*
How do you suppose this figure was constructed? What purpose might it have?

The Future of Tourism in Peru

In spite of its potential, the future of tourism in Peru faces some serious challenges. A particularly important one is the political instability in the country. The discrimination against the Indian and *mestizo* majority by the white upper class has caused deep resentment. This resentment is made much stronger by the wide discrepancies in incomes earned by the different racial groups. Descendents of the Spanish colonists continue to hold much of the wealth of the country, while the majority of the Indians live in absolute poverty. In recent years, **leftist guerillas** — those who hold a socialist or communist philosophy — have used this resentment to build support for revolutions against the government. One group in particular, the Shining Path, has used violent tactics to disrupt the government. While terrorism is not directed at tourists, it does cause fear and loss of life and property.

A second difficult problem to overcome is poverty. Tourists are most likely to encounter poverty on the street. Children with their hands extended begging for coins are hard to ignore, especially since they are uncommon in the countries from which tourists come. Another way in which poverty affects tourists is in high crime rates. Common street theft and minor crimes are annoyances vacationers prefer to do without. In Peru, poverty is widespread and will only be solved by serious economic development and careful planning. Improvements have been hampered by the national **foreign debt** (the amount of money owed to other countries) that threatens to strangle the economy of the country. In 1989, Peru owed other countries a total of US$18.6 billion.

Figure 6.10b *Arequipa*

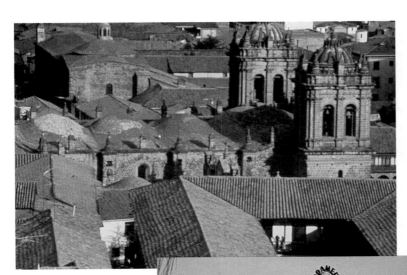

Figure 6.10a *Cuzco*

Figure 6.10c *Lima*

Cities are a major tourist attraction in Peru.

In Peru, as in other developing countries, farmers in remote areas find the most reliable income can be obtained from growing coca for the illegal drug trade. Commercial crops, such as coffee, cotton, and sugar cane, are susceptible to fluctuations in the world market; farmers' incomes can be drastically reduced if prices fall. On the other hand, there is a strong and growing demand for cocaine, the illegal drug obtained from the coca plant. Peru is the world's largest grower of coca, although most of the coca that is grown in Peru is shipped to Colombia for processing into cocaine. Although the drug trade does not directly affect tourism, its publicity helps to create a negative world image. If the problem continues, increasing numbers of tourists will choose other destinations.

Figure 6.10 *Consumer Spending as a Percentage of Total Spending in Peru and in the U.S.*

Category of Spending	Peru	United States
Food and Drink	35	13
Clothing	7	6
Housing and Energy	15	20
Household Goods	7	6
Health and Personal Care	4	15
Transportation and Communications	10	15
Other (including leisure)	22	25
GDP PER CAPITA (US$)	880	16 444

Source: World Quality of Life Indicators

What categories make up the largest part of people's spending? What do you think are the reasons for this?

CHECKBACK AND APPLY

16. **a)** List the economic problems that Peru faces.

b) Which of these problems do you consider most serious? Rank the problems according to their seriousness. Defend your ranking.

c) In what ways will these economic problems affect tourism in the country?

17. The problems discussed in this section are all related to the causes or effects of poverty. How would you solve the problem of poverty in Peru? Write a one-page statement outlining your plan.

18. Figure 6.10 compares consumer spending by Peruvians and Americans. On the surface, some of the categories (for example, spending on household goods) show similarities.

a) List some typical household goods that would be bought by consumers in each of the two countries.

b) What effect would the amount of GDP per capita have on consumer spending?

c) Why would there be such large differences in spending on food and drink?

d) Suppose the government of Peru subsidized farmers so that food costs were cut in half. What effect would that have on spending in other areas? Would this be a sensible course of action for the government to take?

Chapter Review

- Indian peoples have occupied the area that is Peru for thousands of years. Some groups, including the Incas, built great civilizations.

- Spanish adventurers conquered the Incas and established a new and unequal society, through which they controlled the economic life of the colony.

- After the country achieved independence, the people of Spanish origin continued to hold power over the Indian peoples, almost all of whom lived in great poverty.

- The deep split between the peoples of Peru remains today, and causes considerable resentment among the Indians and the *mestizos*, people of mixed blood.

- The magnificent remains of the Indian societies, particularly the Incas, attract researchers and the curious to Peru. The Spanish colonial architecture and the way of life also draw tourists to the country.

- The tourist potential is limited by problems that stem from the cultural split in the country and the overwhelming poverty of the people of Indian background.

 ## Vocabulary Review

foreign debt   mestizos
historic cultural region   terraces
leftist guerillas

 ## Thinking About...

PERU AS A HISTORIC CULTURAL REGION

1. Summarize the characteristics that allow Peru to be identified as a historic cultural region.

2. **a)** "Peru's physical geography is both one of its most valuable resources and one of its greatest challenges." Explain what this means.

   b) The Incas built their empire across and beyond Peru. What does this fact show about their leadership, administration, and engineering skills?

   c) In what ways does the tourist industry in Peru take advantage of the physical environment of the country?

3. Many important sites of the Inca civilization and the colonial period have been damaged by robbers who steal artefacts and by tourists who take home souvenirs. What are some measures that the Peruvian government could take to protect these historic cultural sites?

4. **a)** What parallels can be drawn between the Indian situation in Peru and the Indian situation in Canada? In what ways are the situations quite different?

   b) Develop a plan to reduce the level of racial discrimination and to equalize opportunities for minorities in both Peru and Canada.

5. Suppose you wanted to take advantage of some of the government programs to develop tourism in Peru. Design a tourist facility based on the historic cultural attractions of Machu Picchu, targeted for North American tourists. Include in your written plan:

   - buildings that would be needed
   - the infrastructure that would be necessary
   - jobs that would be created
   - problems that might be encountered

LOOKING BACK AT PERU

6. In an attempt to deal with economic problems, including a huge foreign debt, the government of Peru has lowered the value of its currency compared to other countries. Here is the number of intis that one U.S. dollar would buy:

1985	10.97 intis
1986	13.95
1987	16.84
1988	128.83
December 1989	5261.40
June 1990	7289.00

a) For the average citizen of Peru, what are the effects of lowering the value of the currency?

b) Does a lower value for the intis improve or worsen the picture for tourism in Peru? Explain your answer.

c) Suppose you were a tourist leaving for Peru. How would you deal with the rapidly falling value of the intis? How could you benefit from the situation?

7. a) List some of the problems that Peru faces as a country. Write a brief description of each problem.

b) Which of the problems it faces is the most serious? Explain your answer clearly.

c) What do you think other developing countries can learn from Peru's experiences? How could Peru share its ideas?

Atlas Activities

1. Refer to the maps on pages A30 and A31.

a) Describe the location of Peru relative to the major landform regions of South America.

b) What is the most dominant feature of Peru's landform patterns? How might this feature influence people's use of the land?

c) Explain the relationship between the vegetation pattern and landforms in Peru.

d) What appears to be the relationship between the distribution of population in Peru and Peru's physiography?

2. The Earth is divided up physically into a number of "plates", large sections of the Earth's crust. These plates are constantly in motion, although the motion is very slow and difficult to detect. It is most noticeable where plates meet. At these boundaries, earthquakes and volcanos occur and mountain ranges are created.

a) Using the map on page A10, describe the location of the Andes Mountains relative to plate boundaries.

b) What other mountain ranges appear to be connected to the Andes chain?

c) What signs of instability in the Earth's crust appear in Peru?

Further Explorations

1. Select an issue that Peru must deal with in the coming years. Conduct research to find out:

- the parts of the country it affects most
- the causes of the problem
- possible solutions
- the effects of the issue on tourism

Prepare either a written report or a ten-minute presentation on the issue.

2. Compare the Caribbean cultural region to the historic cultural region of Peru. Some topics you should include in your comparison are:

- important cultural influences
- time periods of greatest importance
- cultural similarities and differences
- problems and solutions

What other topics would make good points for comparison?

3. Imagine that you are the person responsible for giving Canadian international assistance (foreign aid). Peru requires help in developing tourism in the country. Some questions that you ask yourself include:

a) What types of tourist facilities are most likely to improve the situation for tourists?

b) What types of tourist facilities are most likely to improve the situation for local people?

c) In what parts of the country should most of the new development take place?

d) Would the assistance be better given to another sector of the economy, such as mining or industry?

e) How will the country's current economic and social problems affect the success of the foreign aid in helping the people?

f) What benefits will Canada realize from giving assistance?

g) Are there other countries that would benefit more from Canada's help?

Write a two-page report in which you agree or decline to give the Peruvians international assistance. Justify your decision by referring to the points covered in your answers to the questions above.

4. Archaeology, the study of prehistory and ancient periods of history based on the examination of their physical remains, has fascinated people over the centuries. Some important archaeological events are noted on the time line below.

By researching one of the events on the time line, or another of your choice, find out what is involved in an archaeological study.

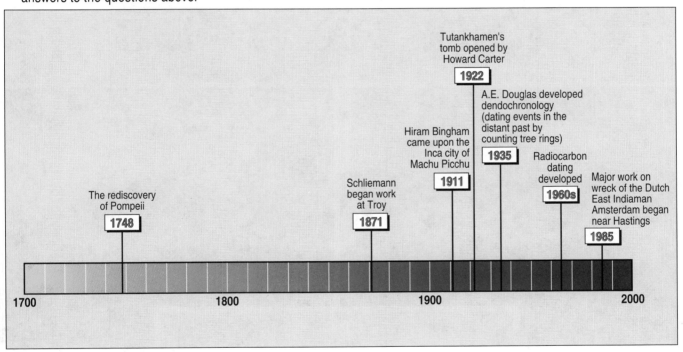

7

Kenya
A PHYSICAL REGION

Figure 7.a *A Typical Village in the Kenyan Highlands*

Figure 7.b *A Sisal Plantation in the Great Rift Valley*

Figure 7.c   *A Kenyan animal trainer displays a trained monkey.*

Figure 7.d *A View of Downtown Nairobi*

Figure 7.e *Zebras on the Savanna*

STARTING OUT

1. By studying the photos on these pages, what can you tell about Kenya's physical environment — its vegetation, climate, and landforms?
2. Kenya is on the equator. What effect do you think this has on its climate?
3. What do these photos tell you about how the people in Kenya live? What are their homes like?
4. What features of Kenya shown in these photos would appeal to you as a tourist?

LEARNING DESTINATIONS

By the end of this chapter, you should be able to:
- understand what constitutes a physical region;
- describe the landforms, climate, and vegetation of Kenya's savanna ecosystem;
- evaluate the tourist potential of the country;
- understand the importance of some of the key developments taking place in Kenya.

Introduction

Chapter 1 described how you can use regions as tools for understanding the world. We need such tools to simplify and organize the vast amount of data that we constantly encounter. Regions allow us to concentrate on one specific area at a time. If the characteristics that we use to define a region are appropriate, we can better understand that region and draw informed conclusions about it. We define Kenya, the region of study in this chapter, by its physical characteristics — its landforms, climate, and vegetation.

The equator runs across Kenya. This factor alone gives clues that we can use to investigate its physical environment. Another important factor is its location on the Indian Ocean. These factors, along with other influences, help to create remarkably interesting physical characteristics.

CHECKBACK AND APPLY

1. Find Kenya on the map on pages A22-A23.

   a) What lines of latitude and longitude are closest to Nairobi?

   b) Name the countries that border on Kenya.

2. Give one advantage and one disadvantage of Kenya's location for tourists from France, Japan, and your area.

3. Look back again at the photos on the opening pages of this chapter. What images do you already have of Kenya? In groups, brainstorm lists of words to describe Kenya. Use these topics as headings for your lists:

   - climate
   - wildlife
   - people
   - ways of life

   Look back at your lists and think about their accuracy as you work your way through this regional study of Kenya.

Kenya's Ecosystem

Much of Kenya is a **savanna ecosystem**. Savanna means grassland. An ecosystem is a system made up of all the living things (plants and animals) and the nonliving things (landforms, climate, and soils) interacting in a given place. A savanna ecosystem is distinguished from other ecosystems by the following characteristics:

- warm the year round, with hot seasons
- distinct dry and wet seasons
- grasses dominate, but low, thorny trees also grow
- great numbers of hooved, grazing animals and **carnivores** (meat-eating animals)

All parts of the savanna ecosystem are **interdependent**. That means that if one element changes, such as the number of large carnivores, then all the other elements will change.

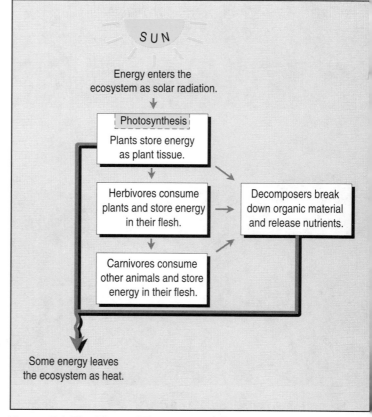

Figure 7.1 *The Flow of Energy in an Ecosystem*
All life in an ecosystem depends on plants and the process of photosynthesis.

An important way in which all the elements of an ecosystem are interdependent is through the flow of energy. This takes place in a series of steps called a **food chain**. Energy comes from the sun. Plants use the sun's energy to make carbohydrates. This is called **photosynthesis**. Plant-eating animals (**herbivores**) then consume the plant tissue where the energy is stored. Meat-eating animals consume other animals, thereby taking the energy that is stored in the flesh. Throughout this food chain, **decomposers**, such as bacteria, are at work. Decomposers feed on dead organisms from all levels of the food chain. They break down organic material, releasing the nutrients and making them available again for plant growth. Figure 7.1 illustrates this flow of energy within an ecosystem.

In Kenya's savanna ecosystem, the dominant **producers** (makers of carbohydrates) are grasses, although the shrubs and trees that grow in scattered locations also produce some carbohydrates. Grazing animals — elephants, zebras, giraffes, wildebeests, and the like — thrive on the savanna grasses in the open, park-like setting. Lions, humans, and other large carnivores feed on these herbivores. Humans are part of this ecosystem as consumers of both plants and animals.

The most lush part of the Kenyan savanna occurs in the southern portion of the country, south of the Tana River. (See map on page A23.) Trees here are more abundant and the grasses are taller. The grasses in the northern part of the country are sparse and there are only shrubs and brush trees.

During the rainy season, between November and May, the streams overflow, the lakes fill, and the land turns green. During the dry season, the hot sun dries up the water, the land becomes dusty, and the grasses and many animals die.

CHECKBACK AND APPLY

4. a) In your own words, or in a drawing, describe how a savanna ecosystem works.

   b) About 15 million square kilometres of the Earth are classified as savanna ecosystems. Refer to the maps on pages A12 and A13. Name one other part of the world that you think has a savanna ecosystem similar to that in Kenya. Give reasons for your choice.

5. Describe three specific ways in which human activity can harm the operation of an ecosystem. State whether each activity has a long-term or short-term impact.

6. Why do you think the grasses in the southern part of Kenya are more lush than those in the northern part?

7. a) List several tourist attractions of the Kenyan savanna ecosystem.

   b) Why might people want to visit this ecosystem? What might they learn?

   c) Write a tourist advertisement for the Kenyan savanna ecosystem. What would you feature to attract tourists? Where might you place your ad — in a newspaper or magazine? on TV or radio? on a billboard? Why?

Figure 7.2 *The Kenyan Savanna*
The grasses of the Kenyan savanna are hardy enough to survive dry periods, fires, and grazing by wildlife. An extensive root system is one of the keys to their survival. The acacia trees are also well adapted to the hot climate and irregular rainfall.

The Climate of Kenya

So far in this chapter we have looked at the living elements in Kenya's savanna ecosystem. But ecosystems are made up of both living and nonliving elements. The living environment of Kenya is greatly influenced by its climate, which varies greatly throughout the country. The rainfall map (Figure 7.3) and the climate statistics given in Figure 7.4 illustrate this. In the southeast and western parts of the country, rainfall exceeds 1000 mm annually, while large parts of northern and central Kenya are quite arid, with annual rainfall totals of less than 250 mm. This explains the sparse grasslands in the northern half of the country and the more lush conditions in the southern half.

Three factors are important in shaping the climate of this region: landforms, the Indian Ocean, and the Earth's tilt.

LANDFORMS

Landforms play a major role in the climate. Notice from Figure 7.5 that the land rises steadily as you move westward from the Indian Ocean. Nairobi, about 400 km from the ocean, is at an elevation of 1820 m, while Mount Kenya, just over 100 km to the north of Nairobi, peaks at 5199 m. Places at higher elevation have cooler temperatures than those closer to sea level.

The differences in temperature because of elevation produce quite distinct vegetation zones. One traveller who journeyed up Mount Kenya gave this description:

Figure 7.3 _Annual Rainfall in Kenya_
Notice the wide variations in rainfall in Kenya. Name three factors that you think could create these rainfall patterns.

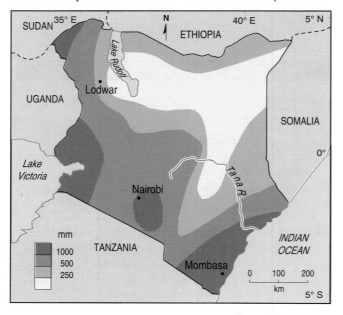

Figure 7.4 _Climate Statistics for Nairobi, Mombasa, and Lodwar_
The top row for each city gives the average maximum temperature for each month in degrees Celsius.
The middle row for each city gives the average minimum temperature for each month in degrees Celsius.
The bottom row shows the average number of days with measurable precipitation.
Which of these three places would you, as a tourist, prefer to visit in May? Explain your choice.

Cities		J	F	M	A	M	J	J	A	S	O	N	D
Nairobi	Max. Temp. (°C)	25	26	25	24	22	21	21	21	24	24	23	23
(1820 m, 1°S, 36°E)	Min. Temp. (°C)	12	13	14	14	13	12	11	11	11	13	13	13
	Precip. (days)	5	6	11	16	17	9	6	7	6	8	15	11
Mombasa	Max. Temp. (°C)	31	31	31	30	28	28	27	27	28	29	29	30
(16 m, 4°S, 39°E)	Min. Temp. (°C)	24	24	25	24	23	23	22	22	23	24	24
	Precip. (days)	6	3	7	15	20	15	14	16	14	10	10	9
Lodwar	Max. Temp. (°C)	33	37	37	35	35	34	33	33	35	36	35	35
(506 m, 3°N, 35°E)	Min. Temp. (°C)	22	23	24	24	24	24	23	22	23	24	23	23
	Precip. (days)	0	1	2	4	2	1	1	1	1	1	1	1

"*Mount Kenya is the second-highest mountain in Africa, after Mount Kilimanjaro, and is located on the equator. There is always snow on the top. We drove to the bottom of the mountain and continued from there by foot. It was tough going because the jungle was thick and we were always wet. All we carried were small backpacks with some canned food and small bedrolls. The higher we went, the colder it got and the thinner the air became. As we neared the top, we had to stop frequently to catch our breath and to take a drink from the ice-cold streams that came splashing down the mountainside. Finally, after what seemed forever, but was only three days, we reached the top. Of the original twenty-two who started the trek, only eight made it to the summit. It was cold, but exhilarating, to stand at the top of the mountain and look down. Because of the cold, we stayed there for only a short while and then quickly started down again.*"

Alpesh Bharkhada, a teenager from Mombasa

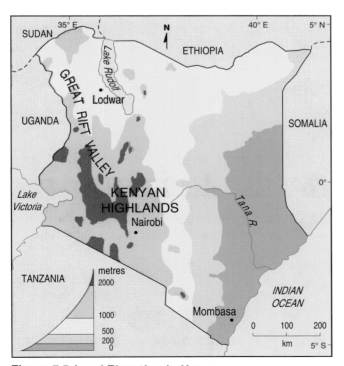

Figure 7.5 *Land Elevation in Kenya*
Using this map, draw a cross-section from Lake Victoria to the Indian Ocean.

THE INDIAN OCEAN

A second important factor that shapes the climate of Kenya is the presence of the Indian Ocean. Winds that cross this large, warm body of water deliver moisture to the land nearby. Mombasa, not far from the ocean, receives over 1000 mm of rain, while Lodwar, deep in the interior of the country, has rain on an average of only 16 days a year. Mountain slopes that face the winds from the Indian Ocean also benefit. The winds cool as they rise over the mountains and the moisture in the air condenses and falls as rain. Because of this, the highland areas in the west of Kenya have annual rainfall totals in excess of 1000 mm.

THE EARTH'S TILT

The seasonal pattern of wet and dry periods results from the relative position of the sun to the equator. In the early months of the year, as the sun moves towards the Tropic of Cancer at 23½°N, the lands in the northern hemisphere are heated more intensely than those to the south. Air above the land is heated, and cooler air flows in from surrounding areas, forcing the lighter, warmer air up. In Kenya, during these months, air flows in from across the Indian Ocean. The **monsoon** winds (winds that change direction according to the season) cross Kenya on their way northward. This gives Nairobi a pronounced wet season during March, April, and May. (See Figure 7.4.)

As the sun appears to move southward towards 23½°S during the second half of the year, the surfaces of the Earth that are most intensely heated are farther southward. This causes the winds in eastern Africa to blow from the northeast, reaching Kenya after having travelled across the deserts of Saudi Arabia and the Horn of Africa. Nairobi has a dry season from June to October. Mombasa does not have such a dry season during this time of the year because it benefits from being close to the Indian Ocean.

8. Refer to Figure 7.4: Climate Statistics for Nairobi, Mombasa, and Lodwar.

   a) How many rainy seasons does Nairobi have? When do they occur?

   b) Does Mombasa have the same rainy season as Nairobi? Why or why not?

   c) Considering the precipitation patterns for both Mombasa and Nairobi, what problems would farmers raising crops in those areas face?

   d) How might the rainfall pattern affect the tourist industry? Explain your answer.

9. **a)** Describe the climate pattern for Lodwar.

   b) What are the advantages and disadvantages of this climate for tourism in Lodwar?

10. What three reasons can you offer to explain the climatic patterns in Kenya?

11. Suppose you were able to visit a rural area near each of the places named in Figure 7.4. Compare the locations and physical environments of the places using a comparison chart similar to the one that follows.

Points of Comparison	Nairobi	Mombasa	Lodwar
latitude			
elevation			
amount of rainfall			
wet and dry seasons			
hottest and coolest months			
type of vegetation			
typical wildlife			

12. **a)** Based on their climates, which place named in Figure 7.4 would you prefer to live in? Why?

    b) What types of clothing would be suitable for visitors to each of the three places during February? during August? Be specific in your recommendations.

The Great Rift Valley

One of the most striking features of the Kenyan landscape is the Great Rift Valley. Running north to south through the country, the valley is part of a larger rift system that runs through much of eastern Africa, including the Red Sea, and eventually links with rift valleys in the Middle East. In some sections, the walls of the valley drop several hundred metres to the floor below and offer spectacular vistas that have helped make Kenya a popular African tourist destination. The Great Rift Valley also contains some of the most productive farmland in Kenya.

The Great Rift Valley has formed because of shifts on the Earth's surface that occurred 10-20 million years ago. The Earth's surface is made up of a number of plates that are moving in different directions at different speeds.

Two plates run the length of eastern Africa. These plates are moving apart, and great cracks are forming in the Earth's crust. Parts of the Earth's surface between these cracks drop down, creating the deep, steep-sided Great Rift Valley.

A number of lakes have formed on the floor of the Great Rift Valley. One of these, Lake Naraisha, is home to over 3 million flamingoes, as well as other wildlife.

13. Consult the map on pages A22-23, which shows the physical features of Africa. Locate the Great Rift Valley.

    a) Which countries does it run through?

    b) Name three lakes and three mountains that are connected to the Great Rift Valley.

14. List two ways in which the Great Rift Valley hinders and two ways in which it helps economic development in Kenya.

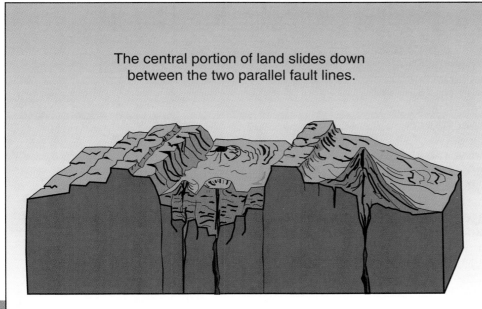

The central portion of land slides down between the two parallel fault lines.

Figure 7.6 *The Formation of a Rift Valley*
A rift valley occurs when two parallel vertical faults occur in the surface of the Earth. The land that lies between those two faults drops down forming the valley. Rift valleys are often regions of earthquake activity. Kenya has, in the past, experienced volcanic activity, although the volcanos are currently dormant (at rest).

Figure 7.7 *The Great Rift Valley in Kenya*
How have the sharp, high sides of the Great Rift Valley affected transportation and other human activity in Kenya?

Tourism in Kenya

Kenya and its savanna ecosystem attract many tourists. Travellers come to experience the subtle beauty of the tree-dotted grasslands, the dramatic changes that come over the land as the rainy season begins, the interplay between the hunter and the hunted on the Kenyan mountainsides, the spectacular slash of the Great Rift Valley. To help preserve the natural beauty of the region, Kenya has established 15 National Parks and 23 public **game reserves** (land set aside by the government for the preservation of wildlife). Altogether, Kenya has over 62 000 km^2 of wildlife sanctuaries. One, Nairobi National Park, is within a few kilometres of the nation's capital city. The Tsavo National Park is over 20 000 km^2 in area. There are strict hunting regulations throughout the country.

Tourists stimulate a good deal of economic activity in Kenya. In 1987, for example, 662 100 tourists visited the country and spent 5840 million shillings (about US$344 million). This industry is a good way for the country to earn income, since visitors spend much more in Kenya than Kenyans spend on tourism. Figure 7.9 compares receipts from tourism and tourist spending by Kenyans.

Spending by tourists creates jobs in Kenya. People are employed in providing services directly to tourists, such as driving tour buses, providing accommodation, and selling souvenirs. Other jobs spin off from the tourist industry, such as building roads to improve access for tourists and supplying food to hotels. In a country with a heavy reliance on agriculture, these additional jobs help to diversify the economy.

Figure 7.9 *Tourist Spending in Kenya*

	Spending by Foreign Tourists in Kenya	Spending by Kenyans on Tourism
Total value (US$)	$344 million	$21 million
% of Kenya's GDP*	5.0	0.3
% of world spending	0.2	0.02

* Gross Domestic Product   Source: *Book of Vital World Statistics*

The leading sources of tourists for Kenya are the United Kingdom and Germany.

CHECKBACK AND APPLY

15. What effects — positive and negative — might tourism have on Kenya's wildlife?

16. a) List some ways in which National Parks and game reserves benefit both the people of Kenya and tourists.

    b) Think of ways in which tourist spending in Kenya benefits more people than just those who come in contact with the travellers. Use a flow diagram or an illustration to show how the benefits spread throughout the economy.

Figure 7.8 *Nakuru National Park*
The National Parks of Kenya were set up to preserve wildlife and areas of outstanding natural beauty. Nakuru National Park, in the Great Rift Valley, has the highest concentration of flamingoes in the world.

Famous Tourist Destinations

Treetops Hotel

A young lady and her husband spent the night of February 5, 1952, in the Treetops Hotel, a remarkable structure perched five metres off the ground in an ancient mgumu tree. They stayed up most of the night watching wild animals, such as elephants, rhinos, wildebeests, and buffalo, coming out of the forest to drink at the pond just below the hotel. The young lady was Princess Elizabeth of England, and while she was at Treetops Hotel, she received news that her father, King George VI, had died. She was to become Queen of England.

Since that time, many famous and not-so-famous visitors have come to the Treetops Hotel for the same reason that Princess Elizabeth did — to see African wildlife up close. It is an experience not easily forgotten. The unique hotel is located near Nyeri, in the central part of Kenya.

The original hotel burnt down, but the mgumu tree in which it was built survived. A new, larger hotel was built, like the original, of wood, with narrow hallways and tiny rooms. Every day, in late afternoon, visitors arrive. Once all the guests have entered, the wooden steps are pulled up. At dusk, they watch the animals that come out to drink at the pond at the foot of the hotel. Guests are advised to keep their windows closed so that baboons and other wildlife will not come in and make off with their possessions.

Lights are turned on to illuminate the watering hole where animals come to drink. Some people stay up all night to watch the animals that wander past the hotel. Occasionally, the hotel will shake as an elephant scratches its back on one of the posts that support the hotel.

Most of the hotel's clientele are wealthy tourists from the industrialized world.

CHECKBACK AND APPLY

17. Treetops Hotel is accessible only to the wealthy. What alternative type of accommodation can you suggest that would allow more people to enjoy the wildlife?

Figure 7.10 *Treetops Hotel*
Treetops Hotel is a different type of hotel that offers an unusual view of African wildlife.

Tourism Issue: Poaching

The widespread **poaching** of African wildlife threatens the future of the tourist industry in countries like Kenya. Poaching is the illegal killing of animals. Elephants are shot for their ivory tusks, which are used for carvings that are sold throughout the world. After the tusks have been taken, the rest of the elephants' bodies are left to rot in the hot African sun. Most poachers in Kenya come from surrounding countries and are well-equipped and heavily armed with submachine guns. They usually work at night in remote areas of game reserves. Due to the difficulty of tracking these poachers and the lack of money and trained personnel to protect the animals, poaching has drastically reduced the number of elephants in Kenya.

A policy recently introduced in Kenya is to shoot poachers on sight if they are caught in the act of killing animals. Government employees who protect wildlife are now equipped with modern technology, including high-powered rifles and airplanes. In 1989, for the first time, the number of elephants killed was drastically reduced and more poachers than elephants were killed.

African wildlife means more than just income and jobs to most Kenyans. It is part of a heritage that has been passed down over the centuries. To see it destroyed is to see part of the culture disappear. Kenyans are determined to end poaching.

The recently adopted policy of **eco-tourism** offers some hope for the preservation of African wildlife. Eco-tourism is designed to allow tourists to visit an area without upsetting the natural environment. A significant portion of the money generated by eco-tourism is used to protect the environment and expand environmentally safe activities. The success of eco-tourism is dependent upon the local residents being involved in planning tourist activities and benefiting economically from them.

Figure 7.11 *The Loss of Elephants in Africa*
Elephants reproduce slowly and reach breeding age between 15 and 20 years of age.
An elephant cow is pregnant for 20 to 21 months before giving birth.
An average elephant eats over 100 kg of vegetation per day.

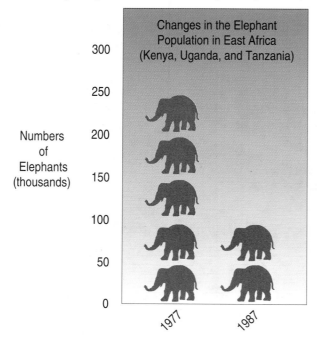

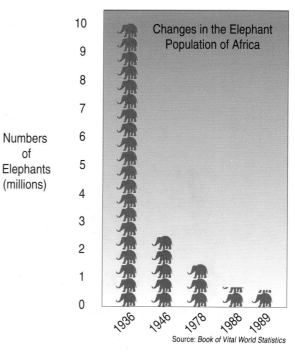

Source: *Book of Vital World Statistics*

CHECKBACK AND APPLY

18. In what ways does poaching threaten the tourist industry in Kenya?

19. Do you agree with the policy of killing poachers if they are caught in the act of killing protected animals? Give reasons for your answer.

20. What are some ways of protecting animals from poachers? Brainstorm a list of ideas that a country like Kenya could use to combat the problem of poaching.

21. In what other regions of the world might a policy of eco-tourism be introduced? Give reasons for your answer.

Figure 7.12 *A Rhinoceros*
Rhinoceroses are slaughtered for the "horn" on the end of their snouts. This valuable "horn" is sold in the Middle East for its medicinal value.

Kenya

THE REGION AT A GLANCE

Area	582 601 km²
Population (1990)	24 393 000
Population density	43.6 persons/km²
Urban population	22%
GDP* per capita (1989)	US$360
Highest elevation	5199 m
Land use	
—arable	4.3%
—permanent pasture	6.6%
—forest and wood	6.4%
—other	82.7%
Annual deforestation (1982-87)	0.8%
Growth in food per person (1977-88)	2.5%
Percent of roads paved	12.3%
Foreign aid (1988)	US$808 million
—per person	US$33.80
Main industries	tourism, agriculture, light manufacturing
Main exports	coffee, tea, cotton, sisal
Main imports	crude petroleum, machinery, vehicles, iron and steel

*Gross Domestic Product

TOURIST GUIDE

- Persons entering Kenya need a valid passport and tourist visa.
- Before travelling to Kenya, get all appropriate inoculations and medicine. There is no vaccine against malaria but pills will prevent you from contracting the disease, which is carried by mosquitoes.
- The tsetse fly is common in areas of low elevation. It carries the sleeping sickness which, if untreated, can be fatal. The disease can be avoided by staying away from swampy areas, using insect repellants, and wearing long sleeves at night.
- It is wise to drink bottled water in outlying areas as only 27 percent of the population has access to safe water. Most large cities have modern water treatment systems, and the water supplies there are considered safe.
- U.S. currency is accepted widely throughout Kenya. Money should be carried in the form of travellers' cheques.
- The two main languages spoken in Kenya are English and Swahili. There are a number of other languages used in the country.

The Future of Tourism in Kenya

Most Kenyans are optimistic that tourism will help the country develop a more stable economy. Like many African countries, Kenya's economy suffers from a dependency on primary industries, particularly agriculture. The economic fortunes of the country swing widely as prices for its important commodities — tea, coffee, sisal (a strong, white fibre used for making rope, twine, etc.), and live-stock — fluctuate on world markets. In addition, agricultural productivity is low: although 80 percent of the workforce is employed in agriculture, agriculture earns only 30 percent of the nation's Gross Domestic Product (GDP). Tourism is seen as a way to stimulate non-agricultural jobs that will ease the unemployment rate, estimated at 30 percent.

Figure 7.13 *Land Use in Kenya*
While grazing is widespread in Kenya, the arid climate limits the crops that can be grown. About 75 percent of the population is concentrated on only 10 percent of the land.

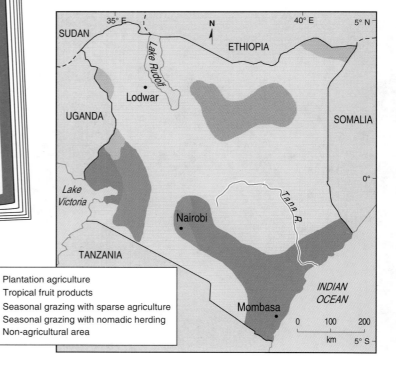

Plantation agriculture
Tropical fruit products
Seasonal grazing with sparse agriculture
Seasonal grazing with nomadic herding
Non-agricultural area

But solving the unemployment problem will be extremely difficult. Kenya has one of the fastest population growth rates in the world. Between 1983 and 1988, the average annual growth rate was 4.9 percent. This compares to growth rates of 0.3 percent for most European countries and a 1.7 percent average for Asian countries. Kenya's population growth is expected to decline to 3.8 percent over the years 1990-2010, but the country will remain one of the fastest growing nations on Earth. Each year, more and more jobs will be needed.

At the present time, over half of all Kenyans are under the age of 15. This situation has been brought about by a declining death rate that has not been matched by a declining birth rate. Improved health care and nutrition mean that fewer children die each year and that people live longer lives. In the years 1985-90, 53.9 children were born each year for every 1000 people in the country. The death rate during those years was 11.9 per 1000 people. By 2010, the country will have an estimated population of 53 million, up from 25 million in 1990.

In addition to needing more jobs, this rapidly growing population also needs land on which to grow food and live. In the 1980s, the amount of food per capita decreased in spite of improved technologies for growing crops. The population simply grew faster than the food supply. Because of this, there is pressure, especially from farmers, to expand outwards and take over more of the savanna for agriculture. Such a move would reduce Kenya's attractiveness for tourists and the potential to stimulate jobs in tourism.

CHECKBACK AND APPLY

22. a) What conditions limit the types of crops that can be grown in Kenya and the productivity of the industry?

b) Based on your knowledge of Kenya, where do you think the best agricultural areas are likely located? Give reasons for your answer.

23. Examine the population pyramid for Kenya shown in Figure 7.14.

a) What does the shape of the pyramid indicate?

b) Estimate the percentage of the people who are between 15 and 64 years of age, the people who make up most of the workforce. Explain the significance of this percentage.

c) Suggest some problems that Kenya faces because of its population profile

d) Suppose the government of Kenya passed a law limiting all married couples to just two children. How would this affect the population growth rate? Explain your answer. What are the human rights implications?

Figure 7.14 *Population Pyramid for Kenya*

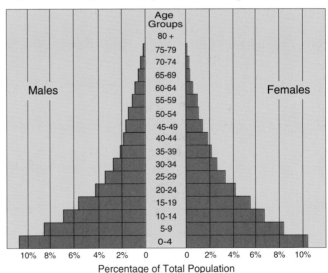

Source: Population Reference Bureau

24. What are some other courses of action, besides expanding onto the savanna, that would lead to increased food production? Brainstorm a list of possible actions that the government could take to feed the increasing population. Choose the best option and write a plan of action suggesting how it might be implemented.

A Critical Challenge: Land Use

Because of Kenya's rapid population growth rate, the use of land in the country is a pressing issue. Some Kenyans are opposed to the preservation of the savanna ecosystem and the establishment of wildlife sanctuaries on their land. They have been there for generations, and the traditions of their ancestors are closely tied to the land. They are, therefore, reluctant to move. Since the soil is highly fertile, it is highly productive, and the local people believe that most, if not all, of the land should be used for farming. Their tribes are rapidly increasing in population and it seems only right to them that they should have use of the land to feed their people rather than cater to a small number of foreign tourists.

Some Kenyans are also concerned that if wildlife reserves are established and they do lose their land, the money made from tourism will not benefit them but would go outside the region (to tourist agencies, airlines, hotel chains, and so on) — perhaps even to other countries. They want Kenya's wildlife resources to be developed for Kenya's benefit. Farmers on more isolated land also complain that the elephants and gazelle are a nuisance. The elephants trample their crops and homes, while the gazelle eat their crops just when they are ready for harvest.

If new wildlife sanctuaries are to be established, the question of land ownership arises. Some land is tribal and is not owned, as we would understand it, by any one person or group of individuals. Rather, land is allocated to individuals according to their need to farm it for food. As a result, it is difficult to compensate the tribes for loss of land. If the local people do not fully support the establishment of a reserve, then it will not operate as efficiently as it might.

A number of interest groups in Kenya are also concerned about the development of game reserves. As discussed earlier, the local people want to preserve their interests. Because of the potentially large deposits of phosphorous and copper discovered on the land, mining companies want to begin explorations. Conservationists, both within Kenya and outside the country, are anxious to see game reserves established as soon as possible. They do not want any competing land uses allowed in reserves. There are also commercial and tourist industry groups who wish to develop the tourist economic potential of reserves without delay. Government officials are anxious to increase tax revenue and earn foreign currency without too great an expenditure of their funds.

CHECKBACK AND APPLY

25. **a)** What land-use conflicts or conflicts over resources exist in your area? in the country as a whole?

b) In what ways are these conflicts similar to and different from those in Kenya? A comparison chart might be useful in completing this comparison.

Chapter Review

- The savanna ecosystem can be used to describe Kenya as a physical region.
- There is a range of savanna conditions within Kenya. This is largely because of the relief of the land and Kenya's location across the equator.
- The Great Rift Valley, a striking feature of the landscape, was formed by a shift in the Earth's surface.
- African wildlife, which attracts many tourists each year, is threatened by poaching.
- Tourism is seen as an industry that will help to create non-agricultural jobs in the country.
- Because of the rapidly growing population, there is pressure on the government to allow National Parks and game reserves to be used for agriculture.

Vocabulary Review

carnivores	interdependent
decomposers	monsoon
eco-tourism	photosynthesis
food chain	poaching
game reserves	producers
herbivores	savanna ecosystem

Thinking About...

KENYA AS A PHYSICAL REGION

1. a) Define the term "physical region".

b) List the characteristics that make Kenya a physical region.

c) Compare a savanna ecosystem to the type of ecosystem that is natural to your area. Some headings you should use in your comparison are:

- landforms
- climate
- vegetation
- wildlife

2. a) What are the most important factors that influence Kenya's climate?

b) In a paragraph, explain the impact of these factors and describe the climate of Kenya.

3. a) List the features that tourists would find attractive in a savanna ecosystem.

b) What types of tourist facilities might be developed in Kenya? Brainstorm some ideas. Present your conclusions in the form of an advertisement for either the print or broadcast media. Where would you place your ad? Why?

4. Suppose you are the Minister for Tourism in the Kenyan government. You want to double the number of tourists to Kenya over the next five years. Prepare a plan of action in which you:

- state your goal
- identify resources (money, people, offices, etc.) that will help you reach your goal
- identify problems and propose solutions (e.g., How will you accommodate an increase in tourists and retain game reserves? How will you ensure that Kenyans benefit economically? How will you minimize any negative effects that tourists may cause?)
- identify ways in which you will measure the changes that your plan brings about

5. Examine Figure 7.15, which shows the population growth for Kenya. Kenya has one of the fastest growing populations in the world. List some advantages and some disadvantages of a rapid population growth to a country such as Kenya.

Figure 7.15 *Population of Kenya*

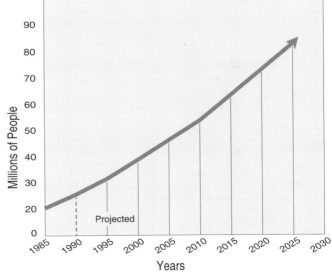

Source: Book of Vital World Statistics

Atlas Activities

1. Refer to pages A22 and A23.

   a) Where are the major deserts in Africa? Give the lines of latitude and longitude of the geographical centre of each.

   b) Give the lines of latitude between which most of the tropical rainforests in Africa are located.

   c) What zone or zones of vegetation lie in-between the deserts and the rainforests?

   d) Name the vegetation zones found in Kenya.

   e) Where is most of the agricultural land, and hence most of the population, of Africa located? What conclusions can you draw from this? Refer to your answers to Questions 1.a) - d) to help you answer this question.

   f) Based on the information contained in this map, as well as in other maps in the Atlas section of the text, what major problems of geography do you think that Africa must deal with if it is to develop further?

2. Study the map of Africa on pages A22 and A23. Using the information given in the map, suggest three ways in which Africa can be identified as a region.

3. Referring to any map in the Atlas section of the text that includes Africa, identify five regions within that continent. Give at least two characteristics of each region you identify and give the latitude and longitude of the geographical centre of each region. You may wish to illustrate the regions on a map of Africa.

4. **a)** Look at the map on pages A2 and A3. What countries other than Kenya does the equator run across?

   b) Put the names of two of these countries in a chart similar to the one shown below. Complete your chart using information from the maps on pages A4-A5, A8-A9, and A12-A13.

	Kenya	Country A	Country B
Population			
Major Landforms			
Agriculture			
Climate			
Vegetation			

   c) Which is most like Kenya in its physical characteristics? Which is least like Kenya in its physical characteristics?

 ## Making Choices

Imagine that you are in charge of developing a game reserve in Kenya. The reserve is to be developed on land that used to contain a great diversity of large and small wildlife and is mainly grassland. Before you can begin to develop this game reserve, you must deal with the land-use issues outlined on page 112. Based on the information you have gained from reading the text and any material that your teacher makes available, answer the following questions. (The game reserve is shown in Figure 7.16.)

1. **a)** What role should the local people have in designing the park?

   b) How should they benefit from the development of the reserve?

   c) How can nearby farms be protected from damage by wildlife?

   d) Should there be multiple use of the reserve? (i.e., Should farming be allowed on the reserve?)

Figure 7.16 *The New Kenyan Game Reserve*
The game reserve that you are to manage is irregular in shape and has only one major road, which crosses the southern edge of the region.

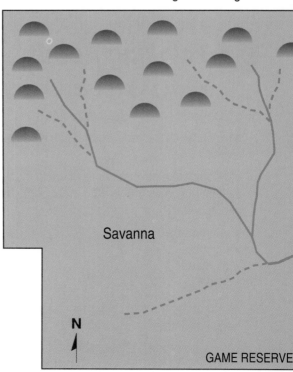

LOOKING BACK AT KENYA

2. What would be an appropriate form of compensation for any loss of land that takes place as a result of the creation of this reserve?

3. As you develop this area for tourists, you are to do the following.

a) Decide on a basic design for the park, including the location of three tourist lodges, access roads, and dirt trails for use by four-wheel-drive vehicles so that visitors can reach the more remote areas of the reserve. Also include accommodation for the people who will help to operate the lodges and the reserve itself.

b) Develop a basic infrastructure, such as electrical supply, water sources, and communications sytems.

c) Decide where you will obtain your food supply from and how you will train farmers to produce the quality of food you need. For example, if you intend to obtain food supplies from local farmers, you must help them to produce the food and to obtain the necessary materials, such as fertilizers, to do so.

d) Define specific tasks for the employees in order that the game reserve operates efficiently and that the wildlife and the entire ecosystem are protected.

e) In order for a game reserve to operate to its maximum potential, it is important to attract tourists from wealthy countries such as Canada. Consider what types of advertising and which media you will use and any package deals you might offer.

Your final plan should include a detailed map of the game reserve, as well as a written outline of how you will set up and operate this reserve.

 Further Explorations

1. Develop a plan for a tourist facility in Kenya that:
- takes advantage of the potential of the savanna ecosystem
- does minimal damage to the physical environment
- employs the maximum number of Kenyans
- accommodates 10 000 tourists annually

Prepare a written report of your plan. Include details of your facility's location and describe how it would operate.

2. In what other region of the world is a savanna type ecosystem found? How is that region similar or different from Kenya?

3. Select two of the following topics and investigate their significance in Kenya. Be sure to include at least one diagram or map as well as a short description. Describe the tourist potential of each.

a) large carnivores

b) extinct volcanos

c) Lake Victoria

d) the beaches of the Indian Ocean

e) pro-natalist population policies

4. Investigate career opportunities related to the preservation of wildlife sanctuaries in a country of your choice. Determine what qualifications are required, what working conditions are like, and the positive and negative aspects of such employment.

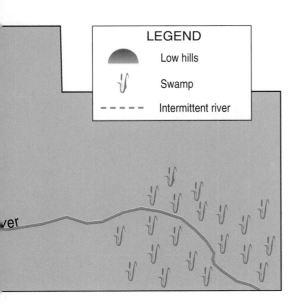

LEGEND

Low hills

Swamp

- - - - - Intermittent river

ATLAS

Contents

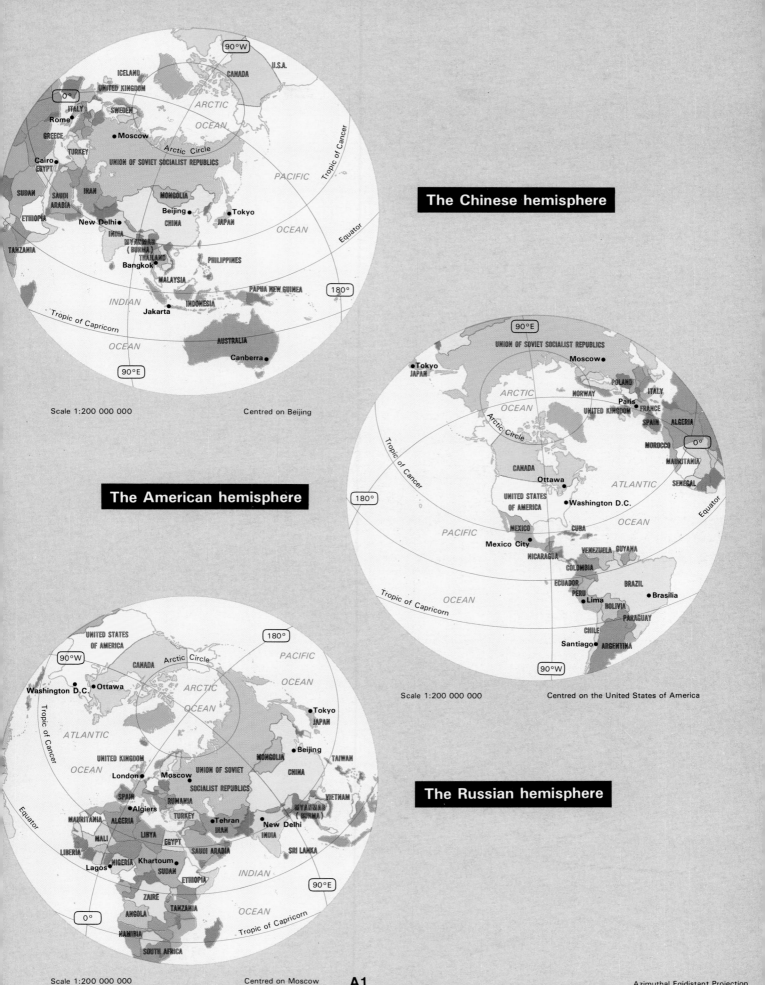

The Chinese hemisphere

Scale 1:200 000 000   Centred on Beijing

The American hemisphere

The Russian hemisphere

Scale 1:200 000 000   Centred on the United States of America

Scale 1:200 000 000   Centred on Moscow

A1

Azimuthal Eqidistant Projection

World— political units

60°W   30°W   0°   30°E   60°E   90°E   120°E

ARC

Greenland
(Denmark)

Svalbard
(Norway)

Arctic Circle

ICELAND

SWEDEN

FINLAND

NORWAY

UNION OF SOVIET SOCIALIST REPUBLICS

60°N

UNITED
KINGDOM

DENMARK

IRELAND

NETHERLANDS
BELGIUM
LUXEMBOURG
SWITZERLAND

GERMANY
POLAND
CZECHOSLOVAKIA
HUNGARY
AUSTRIA
RUMANIA

MONGOLIA

NORTH
KOREA
SOUTH
KOREA

FRANCE

ITALY
YUGOSLAVIA
BULGARIA

ALBANIA

CHINA

Azores
(Portugal)

PORTUGAL

SPAIN

GREECE

TURKEY

AFGHANISTAN

NEPAL

BHUTAN

JAP

Madeira
(Portugal)

TUNISIA

MALTA

CYPRUS
LEBANON
ISRAEL

SYRIA
IRAQ

JORDAN

IRAN

PAKISTAN

BANGLADESH

MYANMAR
(Burma)

LAOS

TAIWAN

Hong Kong
(U.K.)

Canary Islands
(Spain)

MOROCCO

KUWAIT

Tropic of Cancer

ALGERIA

LIBYA

EGYPT

SAUDI
ARABIA

BAHRAIN
QATAR
UNITED
ARAB EMIRATES

OMAN

INDIA

THAILAND

VIETNAM

CAMBODIA

PHILIPPINES

Nort
Mari.
(U.S

MAURITANIA

CAPE VERDE
ISLANDS
SENEGAL
GAMBIA
GUINEA
BISSAU
SIERRA
LEONE
LIBERIA

MALI

NIGER

CHAD

YEMEN

YEMEN
(PEOPLE'S DEMOCRATIC
REPUBLIC)

MICR

BELA

BURKINA
FASO

NIGERIA

GUINEA

IVORY
COAST

GHANA
TOGO
BENIN

SUDAN

DJIBOUTI

SRI LANKA

MALDIVES

BRUNEI
MALAYSIA

SINGAPORE

CAMEROON

CENTRAL
AFRICAN
REPUBLIC

ETHIOPIA

SOMALIA

EQUATORIAL
GUINEA
SAO TOME
AND PRINCIPE

GABON

CONGO

UGANDA

KENYA

0°   Equator

Ascension
(U.K.)

ZAIRE

RWANDA
BURUNDI

TANZANIA

SEYCHELLES

Chagos Archipelago
(U.K.)

INDONESIA

INDIAN

St Helena
(U.K.)

ANGOLA

ZAMBIA

COMOROS

MOZAMBIQUE

MALAWI

Cocos Islands
(Australia)

MADAGASCAR

MAURITIUS
Reunion
(France)

OCEAN

NAMIBIA

ZIMBABWE

BOTSWANA

Tropic of Capricorn

ATLANTIC

SWAZILAND

AUSTRAL

OCEAN

SOUTH AFRICA

LESOTHO

30°S

Tristan da Cunha
(U.K.)

Gough Island
(U.K.)

Kerguelen Islands
(France)

Heard Island
(Australia)

30°W   0°

KILOMETRES
0   700   1400   2100   2800   3500

1 centimetre on the map measures
680 kilometres on the ground

30°E   60°E   A2   90°E   120°E

| 4:00 A.M. | 5:00 A.M. | 6:00 A.M. | 7:00 A.M. | 8:00 A.M. | 9:00 A.M. | 10:00 A.M. | 11:00 A.M. | 12:00 noon | 1:00 P.M. | 2:00 P.M. | 3:00 P.M. |

Tuesday | Monday

180° 150°W 120°W 90°W 60°W 30°W 0°

Greenland
(Denmark)

Arctic Circle

ICELAND

Alaska
(U.S.A.)

60°N

CANADA

PACIFIC

UNITED STATES
OF
AMERICA

ATLANTIC

OCEAN

OCEAN

Azores
(Portugal)

Bermuda
(U.K.)

30°N

MEXICO

BAHAMAS

Tropic of Cancer

Hawaiian Islands
(U.S.A.)

CUBA   DOMINICAN REPUBLIC
JAMAICA   HAITI   Puerto Rico (U.S.A.)
BELIZE   ANTIGUA AND BARBUDA
HONDURAS   ST KITTS-NEVIS
GUATEMALA   DOMINICA
EL SALVADOR NICARAGUA ST VINCENT ST LUCIA
                        BARBADOS
COSTA RICA VENEZUELA GRENADA
PANAMA   TRINIDAD AND TOBAGO

MARSHALL
ISLANDS

COLOMBIA GUYANA
          SURINAM
          French Guiana
          (France)

Equator 0°

URU

KIRIBATI

Galapagos Islands
(Ecuador)

ECUADOR

MON
LANDS

TUVALU

PERU

BRAZIL

ATU

WESTERN
SAMOA

French Polynesia
(France)

American Samoa
(U.S.A.)

BOLIVIA

FIJI

New Caledonia
(France)

TONGA

COOK
ISLANDS

PARAGUAY

Tropic of Capricorn

CHILE

Easter Island
(Chile)

30°S

URUGUAY

NEW ZEALAND

ARGENTINA

Macquarie Island
(Australia)

Falkland
Islands
(U.K.)

South Georgia
(U.K.)

180° 150°W 120°W **A3** 90°W 60°W 30°W

Tuesday | Monday

6:00 7:00 8:00 9:00 10:00 11:00 12:00 1:00 2:00 3:00 4:00 5:00
P.M. P.M. P.M. P.M. P.M.   P.M.   midnight A.M. A.M. A.M. A.M. A.M.

World — population

Uninhabited

Under 1 person per square kilometre

1–10 persons per square kilometre

10–25 persons per square kilometre

25–50 persons per square kilometre

50–100 persons per square kilometre

Over 100 persons per square kilometre

Interrupted Mollweide Homolographic Projection

Scale at 0° latitude (equator)

A4

Non-agricultural areas

Commercial dairy and mixed farming

Specialized horticulture

Commercial grain farming

Crop and livestock farming

Intensive subsistence agriculture (rice dominant)

Mediterranean agriculture

Primitive subsistence agriculture

Intensive subsistence agriculture (no crop dominant)

Nomadic herding

Livestock ranching

Scale at 0° latitude (equator)
1 : 140 000 000; 1 centimetre to 1400 kilometres

Interrupted Mollweide Homolographic Projection

A5

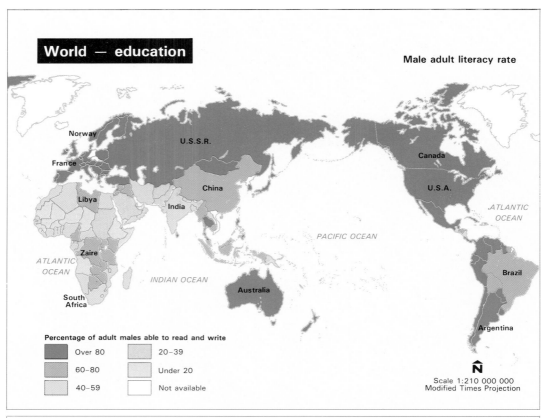

World — education

Male adult literacy rate

Norway
France
U.S.S.R.
China
Libya
India
Zaire
ATLANTIC OCEAN
INDIAN OCEAN
South Africa
Australia
PACIFIC OCEAN
Canada
U.S.A.
ATLANTIC OCEAN
Brazil
Argentina

Percentage of adult males able to read and write

Over 80		20–39
60–80		Under 20
40–59		Not available

N

Scale 1:210 000 000
Modified Times Projection

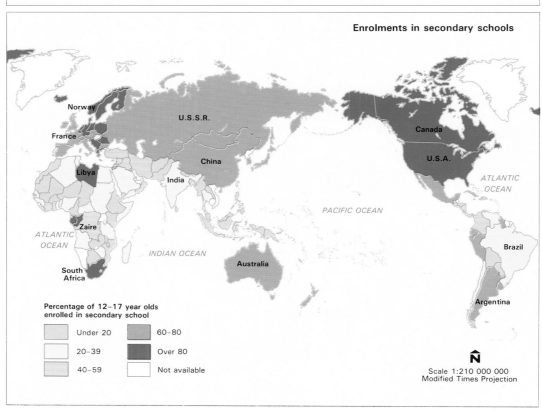

Enrolments in secondary schools

Norway
France
U.S.S.R.
China
Libya
India
Zaire
ATLANTIC OCEAN
INDIAN OCEAN
South Africa
Australia
PACIFIC OCEAN
Canada
U.S.A.
ATLANTIC OCEAN
Brazil
Argentina

Percentage of 12–17 year olds enrolled in secondary school

Under 20		60–80
20–39		Over 80
40–59		Not available

N

Scale 1:210 000 000
Modified Times Projection

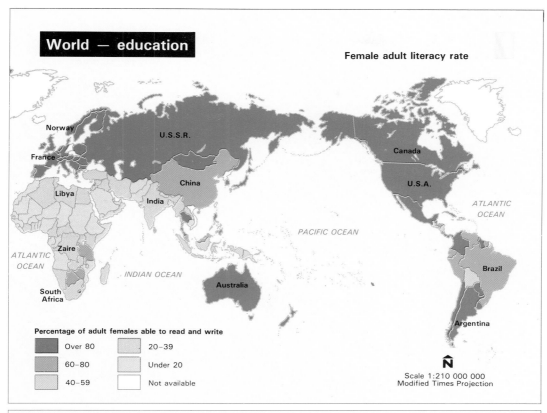

World — education

Female adult literacy rate

Norway
France
U.S.S.R.
China
India
Libya
Zaire
South Africa
Canada
U.S.A.
ATLANTIC OCEAN
PACIFIC OCEAN
INDIAN OCEAN
ATLANTIC OCEAN
Australia
Brazil
Argentina

Percentage of adult females able to read and write

Over 80	20–39
60–80	Under 20
40–59	Not available

N
Scale 1:210 000 000
Modified Times Projection

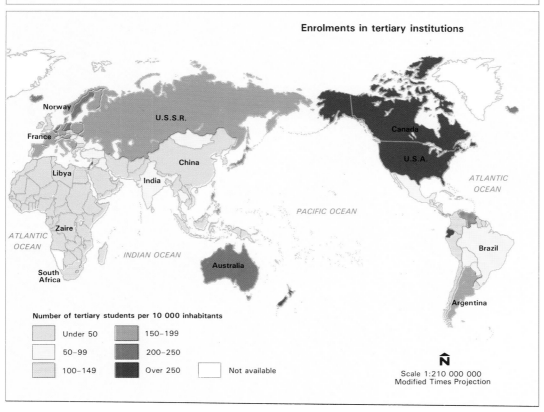

Enrolments in tertiary institutions

Norway
France
U.S.S.R.
China
India
Libya
Zaire
South Africa
Canada
U.S.A.
ATLANTIC OCEAN
PACIFIC OCEAN
INDIAN OCEAN
ATLANTIC OCEAN
Australia
Brazil
Argentina

Number of tertiary students per 10 000 inhabitants

Under 50	150–199
50–99	200–250
100–149	Over 250

Not available

N
Scale 1:210 000 000
Modified Times Projection

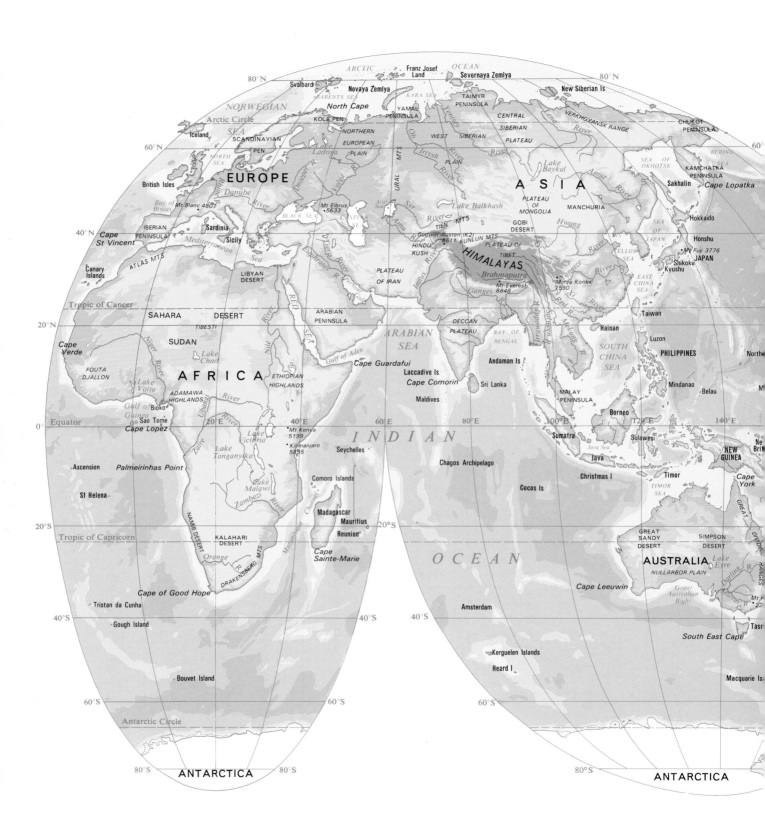

Scale at 0° latitude (equator) 1:90 000 000; 1 centimetre to 900 kilometres
Mollweide Interrupted Homolographic Projection
Heights and depths in metres

A8

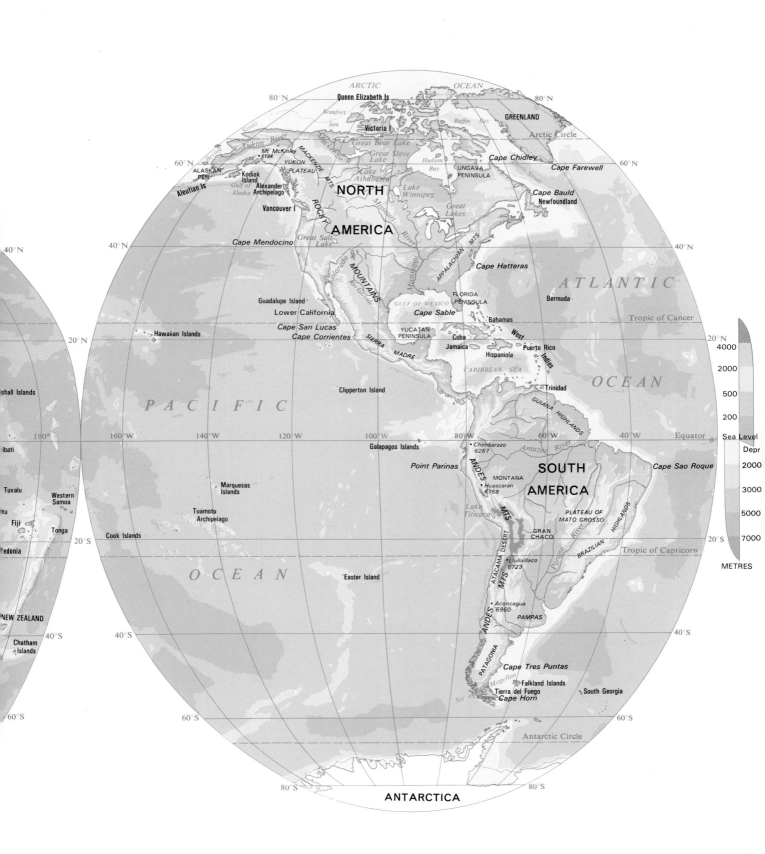

ARCTIC OCEAN

80° N — Queen Elizabeth Is — 80° N

Beaufort Sea

Victoria I

GREENLAND

Arctic Circle

Baffin Bay

60° N — *Bering Strait* — Great Bear Lake — *Hudson Bay* — Cape Chidley — Cape Farewell — 60° N

Yukon River — MACKENZIE MTS — Great Slave Lake

Mt McKinley 6194 — YUKON — Lake Athabasca — UNGAVA PENINSULA — Cape Bauld

ALASKAN PEN — PLATEAU — NORTH — Lake Winnipeg — Newfoundland

Kodiak Island — *Gulf of Alaska* — Alexander Archipelago — ROCKY — AMERICA — Great Lakes

Aleutian Is — Vancouver I

40° N — Cape Mendocino — Great Salt Lake — Missouri River — APPALACHIAN MTS — Cape Hatteras — ATLANTIC — 40° N

Colorado River — MOUNTAINS — Mississippi River

Guadalupe Island — Lower California — *Rio Grande* — FLORIDA PENINSULA — Bermuda

GULF OF MEXICO — Cape Sable — Tropic of Cancer

Hawaiian Islands — Cape San Lucas — SIERRA — YUCATAN PENINSULA — Bahamas

20° N — Cape Corrientes — MADRE — Cuba — West — 20° N

Jamaica — Indies — Puerto Rico — 4000

Hispaniola — 2000

Marshall Islands — CARIBBEAN SEA — OCEAN — 500

PACIFIC — Clipperton Island — Trinidad — 200

Kiribati

Galapagos Islands — GUIANA HIGHLANDS — Equator — Sea Level

180° — 160° W — 140° W — 120° W — 100° W — 80° W — *Amazon River* — 60° W — 40° W — Depr

Tuvalu — Chimborazo 6267 — 2000

Point Parinas — ANDES — SOUTH — Cape Sao Roque — 3000

Western Samoa — MONTANA — AMERICA

Marquesas Islands — Huascaran 6768 — 5000

Fiji — *Tonga* — Tuamotu Archipelago — 7000

Lake Titicaca — PLATEAU OF MATO GROSSO — METRES

Caledonia — Cook Islands — GRAN CHACO — *Parana River* — BRAZILIAN HIGHLANDS

20° S — ATACAMA DESERT — Llullaillaco 6723 — Tropic of Capricorn — 20° S

OCEAN — Easter Island — ANDES MTS

NEW ZEALAND — Aconcagua 6960 — PAMPAS

Chatham Islands — ANDES — Cape Tres Puntas

40° S — 40° W — PATAGONIA — *Magellan* — Falkland Islands — 40° S

Str — Tierra del Fuego — South Georgia

Cape Horn

60° S — 60° S

Antarctic Circle

80° S — ANTARCTICA — 80° S

A9

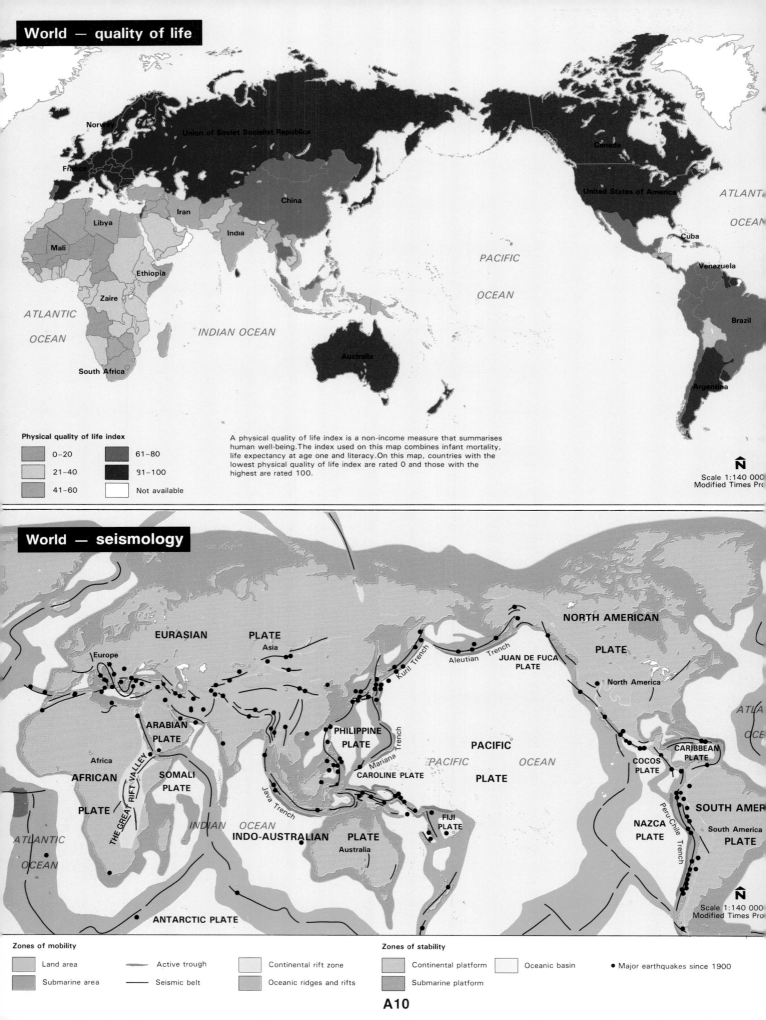

World — quality of life

Norway

France

Union of Soviet Socialist Republics

Canada

ATLANTIC
OCEAN

United States of America

Cuba

Libya

Iran

China

Venezuela

Mali

India

Brazil

Ethiopia

PACIFIC

OCEAN

Zaire

ATLANTIC
OCEAN

INDIAN OCEAN

Argentina

South Africa

Australia

Physical quality of life index

0–20	61–80
21–40	81–100
41–60	Not available

A physical quality of life index is a non-income measure that summarises human well-being. The index used on this map combines infant mortality, life expectancy at age one and literacy. On this map, countries with the lowest physical quality of life index are rated 0 and those with the highest are rated 100.

N

Scale 1:140 000
Modified Times Pro

World — seismology

NORTH AMERICAN

EURASIAN   PLATE

PLATE

Asia

Aleutian   Trench

JUAN DE FUCA
PLATE

Europe

Kuril Trench

North America

ARABIAN
PLATE

PHILIPPINE
PLATE

PACIFIC

CARIBBEAN
PLATE

Africa

Mariana Trench

OCEAN

COCOS
PLATE

AFRICAN

SOMALI
PLATE

CAROLINE PLATE

PACIFIC

PLATE

SOUTH AMER

PLATE

Java Trench

FIJI
PLATE

NAZCA
PLATE

South America

ATLANTIC

INDIAN   OCEAN

INDO-AUSTRALIAN   PLATE

PLATE

OCEAN

Australia

Peru-Chile Trench

THE GREAT RIFT VALLEY

ANTARCTIC PLATE

N

Scale 1:140 000
Modified Times Pro

Zones of mobility

Land area	Active trough
Submarine area	Seismic belt

Continental rift zone	
Oceanic ridges and rifts	

Zones of stability

Continental platform	Oceanic basin
Submarine platform	• Major earthquakes since 1900

A10

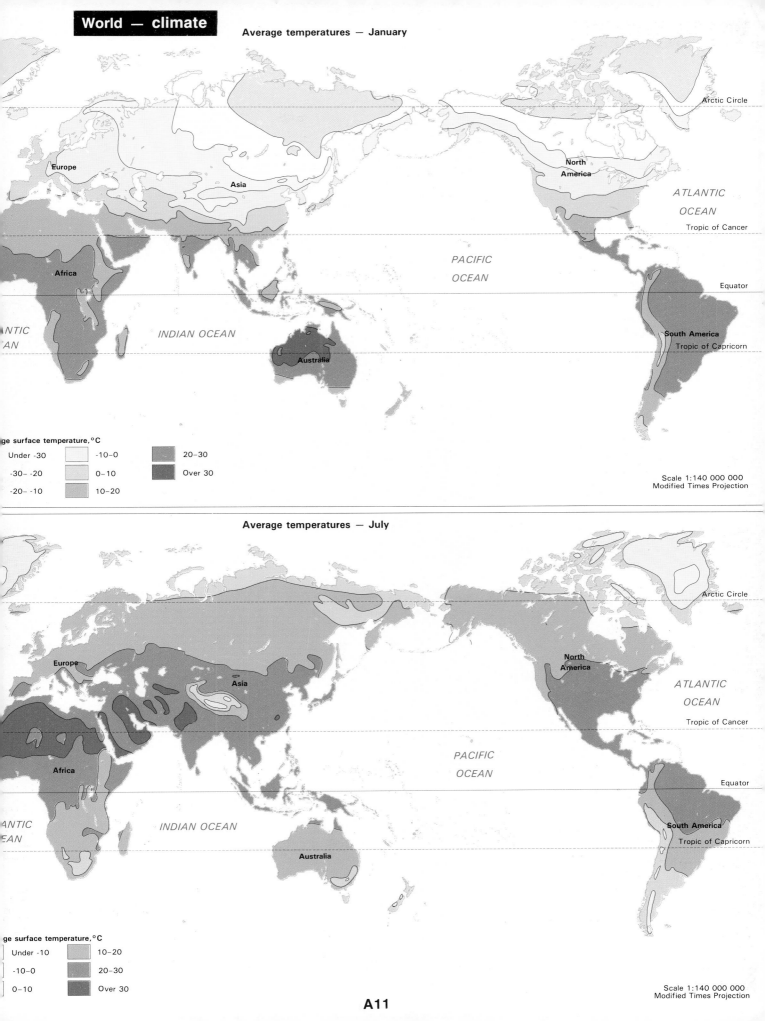

World — climate

Average temperatures — January

Europe
Asia
North America
ATLANTIC OCEAN
Tropic of Cancer
PACIFIC OCEAN
Africa
Equator
INDIAN OCEAN
ANTIC AN
Australia
South America
Tropic of Capricorn
Arctic Circle

ge surface temperature,°C

Under -30
-30– -20
-20– -10
-10–0
0–10
10–20
20–30
Over 30

Scale 1:140 000 000
Modified Times Projection

Average temperatures — July

Europe
Asia
North America
ATLANTIC OCEAN
Tropic of Cancer
PACIFIC OCEAN
Africa
Equator
ANTIC AN
INDIAN OCEAN
Australia
South America
Tropic of Capricorn
Arctic Circle

ge surface temperature,°C

Under -10
-10–0
0–10
10–20
20–30
Over 30

Scale 1:140 000 000
Modified Times Projection

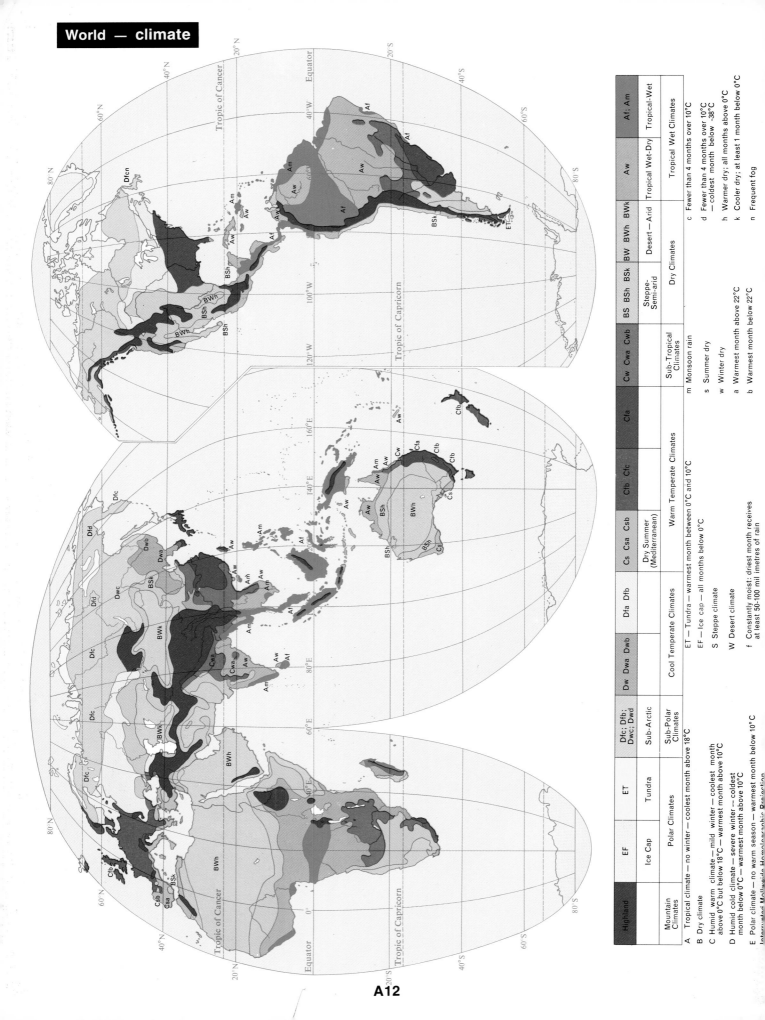

Highland		EF	ET	Dfc; Dfb; Dwc; Dwd	Dw Dwa Dwb	Dfa Dfb	Cs Csa Csb	Cw Cwa Cwb	Cfa	Cfb Cfc	BS BSh BSk	BW BWh BWk	Aw	Af; Am
Mountain Climates		Ice Cap	Tundra	Sub-Arctic		Cool Temperate Climates	Dry Summer (Mediterranean)	Sub-Tropical Climates		Warm Temperate Climates	Steppe-Semi-arid	Desert — Arid	Tropical Wet-Dry	Tropical-Wet
		Polar Climates		Sub-Polar Climates							Dry Climates			Tropical Wet Climates

A — Tropical climate — no winter — coolest month above 18°C

B — Dry climate

C — Humid warm climate — mild winter — coolest month above 0°C but below 18°C — warmest month above 10°C

D — Humid cold climate — severe winter — coldest month below 0°C — warmest month above 10°C

E — Polar climate — no warm season — warmest month below 0°C

ET — Tundra — warmest month between 0°C and 10°C

EF — Ice cap — all months below 0°C

S — Steppe climate

W — Desert climate

f — Constantly moist: driest month receives at least 50-100 millimetres of rain

m — Monsoon rain

s — Summer dry

w — Winter dry

a — Warmest month above 22°C

b — Warmest month below 22°C

c — Fewer than 4 months over 10°C

d — Fewer than 4 months over 10°C — coldest month below -38°C

h — Warmer dry; all months above 0°C

k — Cooler dry; at least 1 month below 0°C

n — Frequent fog

Interrupted Mollweide Homolographic Projection

A12

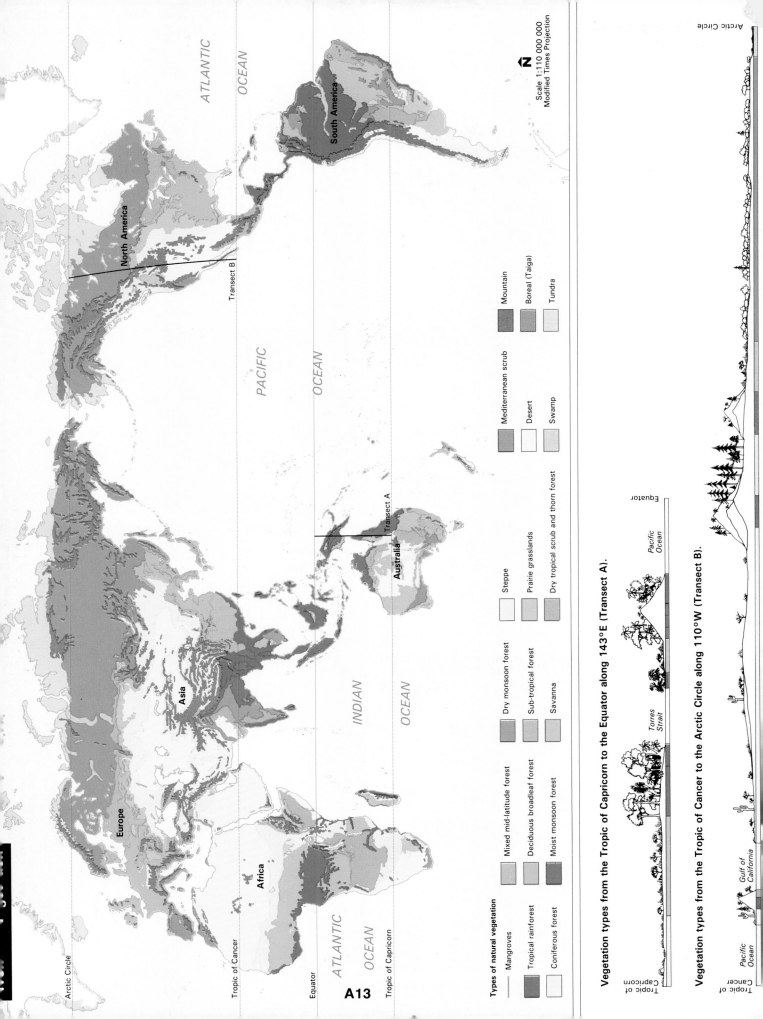

A13

Scale 1:110 000 000
Modified Times Projection

Types of natural vegetation

Mangroves

Tropical rainforest

Coniferous forest

Mixed mid-latitude forest

Deciduous broadleaf forest

Moist monsoon forest

Dry monsoon forest

Sub-tropical forest

Savanna

Steppe

Prairie grasslands

Dry tropical scrub and thorn forest

Mediterranean scrub

Desert

Swamp

Mountain

Boreal (Taiga)

Tundra

Vegetation types from the Tropic of Capricorn to the Equator along 143°E (Transect A).

Vegetation types from the Tropic of Cancer to the Arctic Circle along 110°W (Transect B).

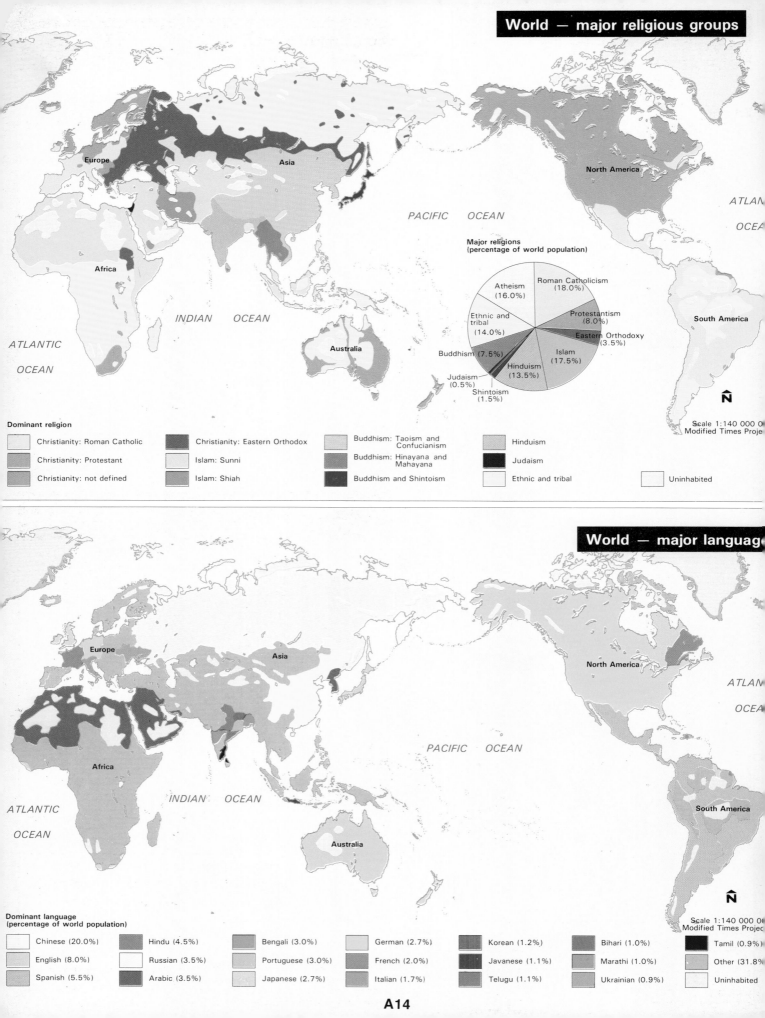

World — major religious groups

Europe

Asia

North America

PACIFIC OCEAN

ATLANTIC OCEAN

Africa

INDIAN OCEAN

South America

Australia

ATLANTIC OCEAN

N̂

Scale 1:140 000 0
Modified Times Proje

Major religions (percentage of world population)

- Roman Catholicism (18.0%)
- Atheism (16.0%)
- Ethnic and tribal (14.0%)
- Protestantism (8.0%)
- Eastern Orthodoxy (3.5%)
- Islam (17.5%)
- Buddhism (7.5%)
- Hinduism (13.5%)
- Judaism (0.5%)
- Shintoism (1.5%)

Dominant religion

- Christianity: Roman Catholic
- Christianity: Protestant
- Christianity: not defined
- Christianity: Eastern Orthodox
- Islam: Sunni
- Islam: Shiah
- Buddhism: Taoism and Confucianism
- Buddhism: Hinayana and Mahayana
- Buddhism and Shintoism
- Hinduism
- Judaism
- Ethnic and tribal
- Uninhabited

World — major language

Europe

Asia

North America

ATLAN OCEA

Africa

PACIFIC OCEAN

INDIAN OCEAN

South America

ATLANTIC OCEAN

Australia

N̂

Scale 1:140 000 0
Modified Times Projec

Dominant language (percentage of world population)

- Chinese (20.0%)
- English (8.0%)
- Spanish (5.5%)
- Hindu (4.5%)
- Russian (3.5%)
- Arabic (3.5%)
- Bengali (3.0%)
- Portuguese (3.0%)
- Japanese (2.7%)
- German (2.7%)
- French (2.0%)
- Italian (1.7%)
- Korean (1.2%)
- Javanese (1.1%)
- Telugu (1.1%)
- Bihari (1.0%)
- Marathi (1.0%)
- Ukrainian (0.9%)
- Tamil (0.9%)
- Other (31.8%)
- Uninhabited

Settlement in the Nile Valley

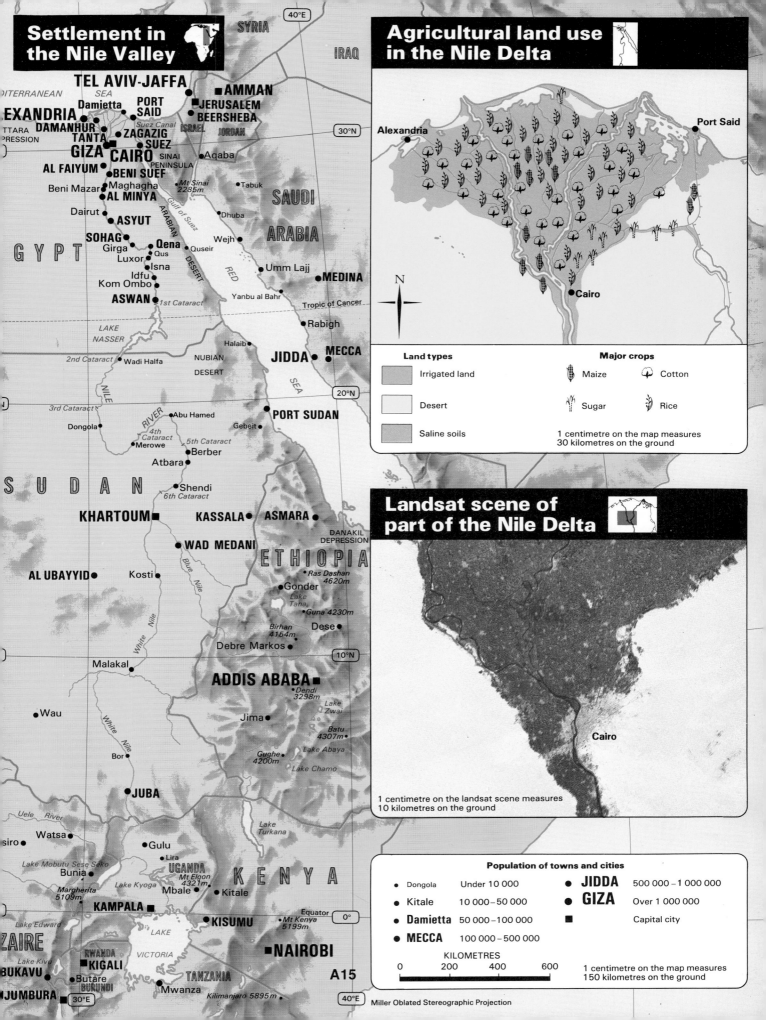

Agricultural land use in the Nile Delta

Land types
- Irrigated land
- Desert
- Saline soils

Major crops
- Maize
- Cotton
- Sugar
- Rice

1 centimetre on the map measures 30 kilometres on the ground

Landsat scene of part of the Nile Delta

1 centimetre on the landsat scene measures 10 kilometres on the ground

Population of towns and cities

• Dongola	Under 10 000	• JIDDA	500 000 – 1 000 000
• Kitale	10 000 – 50 000	● GIZA	Over 1 000 000
● Damietta	50 000 – 100 000	■	Capital city
● MECCA	100 000 – 500 000		

KILOMETRES
0   200   400   600

1 centimetre on the map measures 150 kilometres on the ground

A15

Miller Oblated Stereographic Projection

Landscapes of Asia

ARCTIC OCEAN

NORWEGIAN SEA
NORTH SEA
British Isles
UNITED KINGDOM
■ LONDON
■ PARIS
FRANCE
ITALY
■ ROME
GREECE
■ ATHENS
MEDITERRANEAN SEA
CYPRUS
TURKEY
■ ANKARA
LEBANON
SYRIA
Jerusalem
ISRAEL JORDAN
IRAQ
■ BAGHDAD
KUWAIT
Kuwait
SAUDI ARABIA
BAHRAIN
QATAR
Riyadh ■
Mecca
ARABIAN
PENINSULA
Abu Dhabi
UNITED
ARAB EMIRATES
YEMEN
San'a ■
YEMEN (P.D.R.)
■ Aden
RED SEA
PERSIAN GULF
OMAN
Muscat
GULF OF OMAN
Cape Hadd
ARABIAN SEA

SCANDINAVIAN PENINSULA
NORWAY
SWEDEN
FINLAND
POLAND
WARSAW
HUNGARY
RUMANIA
■ BUCHAREST
BALTIC SEA
Dniester River
BLACK SEA
Mt Elbrus 5633m
CASPIAN SEA
Euphrates River
Tigris River
IRAN
TEHRAN
Damavand Peak 5599m
PLATEAU OF IRAN
HINDU KUSH
KABUL
AFGHANISTAN
Islamabad
PAKISTAN
Indus River
KARACHI
GULF OF OMAN

NORTH CAPE
North Cape
KOLA PENINSULA
LENINGRAD
NORTHERN EUROPEAN PLAIN
Andropov Reservoir
MOSCOW
GORKI
PERM
Volga River
Ural River
CHELYABINSK
1894m Narodnaya
URAL MOUNTAINS
ARAL SEA
Syr River
Amu River
TASHKENT
Communism Peak 7495m
Godwin Austen (K2) 8611m
KUNLUN
Muztag Tower 7723m
PLATEAU OF TIBET
NEW DELHI
Ganges River
NEPAL
Katmandu
Mt Everest 8848m
BHUTAN
Thimphu
BANGLADESH
DACCA
INDIA
DECCAN PLATEAU
WESTERN GHATS
EASTERN GHATS
Krishna River
BOMBAY
MADRAS
Laccadive Islands
Cape Comorin
Colombo ■
SRI LANKA
Dondra Head
Male ■
MALDIVES
0° Equator
INDIAN OCEAN

ARCTIC OCEAN
Svalbard
Franz Josef Land
BARENTS SEA
Novaya Zemlya
KARA SEA
YAMAL PENINSULA
Severnaya Zemlya
Novaya Zemlya
TAIMYR PENINSULA
LAPTEV SEA
New Siberian Islands
EAST SIBERIAN SEA
GULF OF OB
Lower Tunguska River
Yenisey River
Irtysh River
UNION OF SOVIET SOCIALIST REPUBLICS
NOVOSIBIRSK
Lake Zaisan
Lake Balkhash
Pobeda Peak 7439m
Urumqi
GOBI DESERT
Bratsk Reservoir
Irkutsk
Lake Baykal
Ulan Bator
MONGOLIA
PLATEAU OF MONGOLIA
YABLONOVYY RANGE
Aldan River
Yakutsk
Kerulen River
BEIJING
CHINA
Hwang River
CHENGDU
Minya Konka 7590m
Yangtze River
HIMALAYAS
Brahmaputra River
MYANMAR (BURMA)
GUANGZHO
Hanoi ■
LAOS
Vientiane ■
THAILAND
RANGOON
BANGKOK
VIET
CAMBODIA
Phnom Pe
HO CHI MINH
GULF OF THAILAND
Point Baibun
BAY OF BENGAL
Andaman Islands
MALAY PENINSULA
MALAY
Kuala Lumpur
SINGAPORE
SUMATRA
JAKARTA JAV

Major vegetation and land use types

- Tropical rainforest
- Monsoon forest
- Other tall forest
- Dry tropical scrub
- Savanna
- Grassland
- Mediterranean vegetation
- Mangroves
- Swamp
- Mountain vegetation
- Tundra
- Desert
- Agricultural land

Major minerals

□ Aluminium	☆ Mercury
△ Chromium	+ Nickel
✳ Coal	⊙ Oil
⊗ Copper	★ Silver
⊟ Gas	⊖ Tin
◇ Gold	⬦ Tungsten
⊙ Iron	△ Uranium
▲ Lead	⊞ Zinc
◮ Manganese	

0   400   800   1200   1600 km

Scale 1:39 000 000

Lambert Azimuthal Equal Area Projection

A16

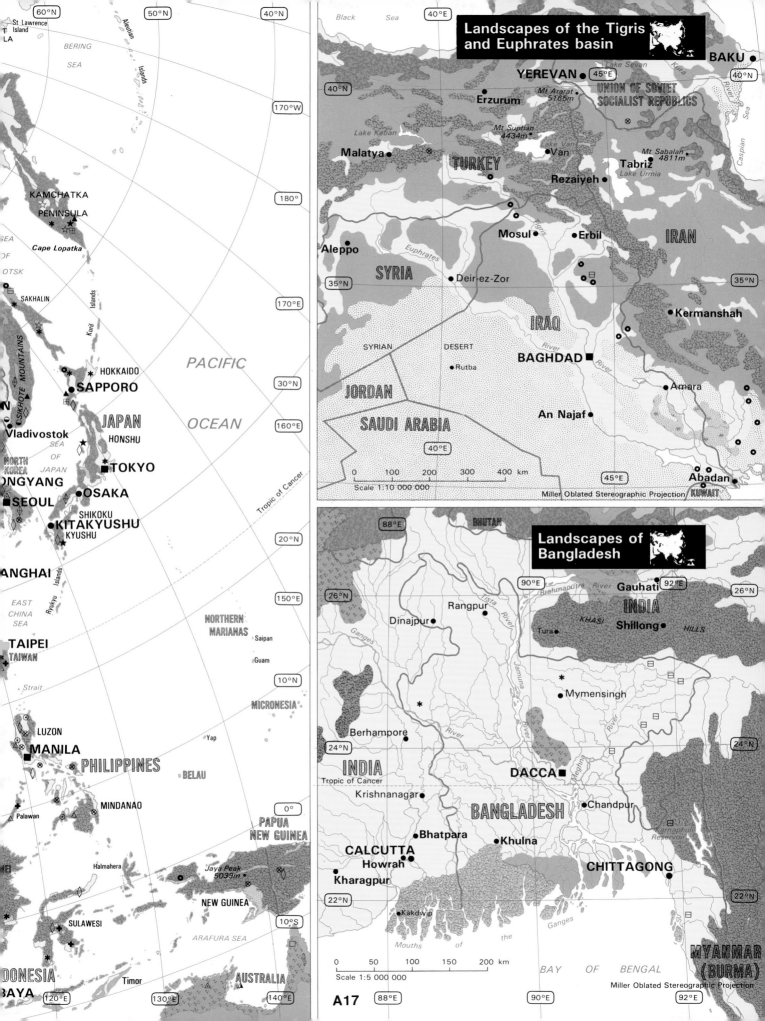

Landscapes of the Tigris and Euphrates basin

Black Sea
40°E
40°N
45°E
40°N
BAKU
YEREVAN
Kura River
Caspian Sea
UNION OF SOVIET SOCIALIST REPUBLICS
Erzurum
Mt Ararat 5165m
Lake Seven
Malatya
Lake Keban
Mt Suphan 4434m
Van
Mt Sabalan 4811m
Tabriz
TURKEY
Lake Van
Lake Urmia
Rezaiyeh
Tigris
Aleppo
Mosul
Erbil
IRAN
Euphrates
SYRIA
Deir-ez-Zor
35°N
35°N
Kermanshah
IRAQ
River
SYRIAN
DESERT
BAGHDAD
River
JORDAN
Rutba
Amara
SAUDI ARABIA
An Najaf
40°E
45°E
Abadan
0   100   200   300   400 km
Scale 1:10 000 000
KUWAIT
Miller Oblated Stereographic Projection

Landscapes of Bangladesh

88°E
BHUTAN
90°E
92°E
Gauhati
26°N
26°N
Tista River
Brahmaputre River
INDIA
Rangpur
Dinajpur
KHASI
Shillong
HILLS
Ganges
Tura
Jamuna
*
Mymensingh
*
River
Berhampore
24°N
24°N
Meghna
INDIA
River
DACCA
Tropic of Cancer
BANGLADESH
Krishnanagar
Chandpur
Karnaphuli Reservoir
Bhatpara
Khulna
CALCUTTA
Howrah
CHITTAGONG
Kharagpur
22°N
22°N
Kakdwip
Ganges
0   50   100   150   200 km
the
Scale 1:5 000 000
Mouths   of
MYANMAR (BURMA)
A17
88°E
BAY   OF   BENGAL
90°E
92°E
Miller Oblated Stereographic Projection

St Lawrence Island
60°N
50°N
40°N
LA
BERING SEA
170°W
KAMCHATKA PENINSULA
Aleutian Islands
Cape Lopatka
180°
SEA OF OTSK
SAKHALIN
170°E
Kuril Islands
SIKHOTE MOUNTAINS
PACIFIC
HOKKAIDO
SAPPORO
30°N
JAPAN
OCEAN
160°E
HONSHU
Vladivostok
SEA OF JAPAN
NORTH KOREA
TOKYO
NGYANG
OSAKA
SEOUL
SHIKOKU
KITAKYUSHU
20°N
KYUSHU
Tropic of Cancer
NORTHERN MARIANAS
Saipan
150°E
ANGHAI
Guam
Ryukyu Islands
EAST CHINA SEA
10°N
TAIPEI
TAIWAN
Yap
MICRONESIA
140°E
Strait
LUZON
MANILA
BELAU
PHILIPPINES
0°
MINDANAO
Palawan
PAPUA NEW GUINEA
Halmahera
Jaya Peak 5039m
10°S
NEW GUINEA
SULAWESI
DONESIA
Timor
ARAFURA SEA
AUSTRALIA
BAYA
120°E
130°E
140°E

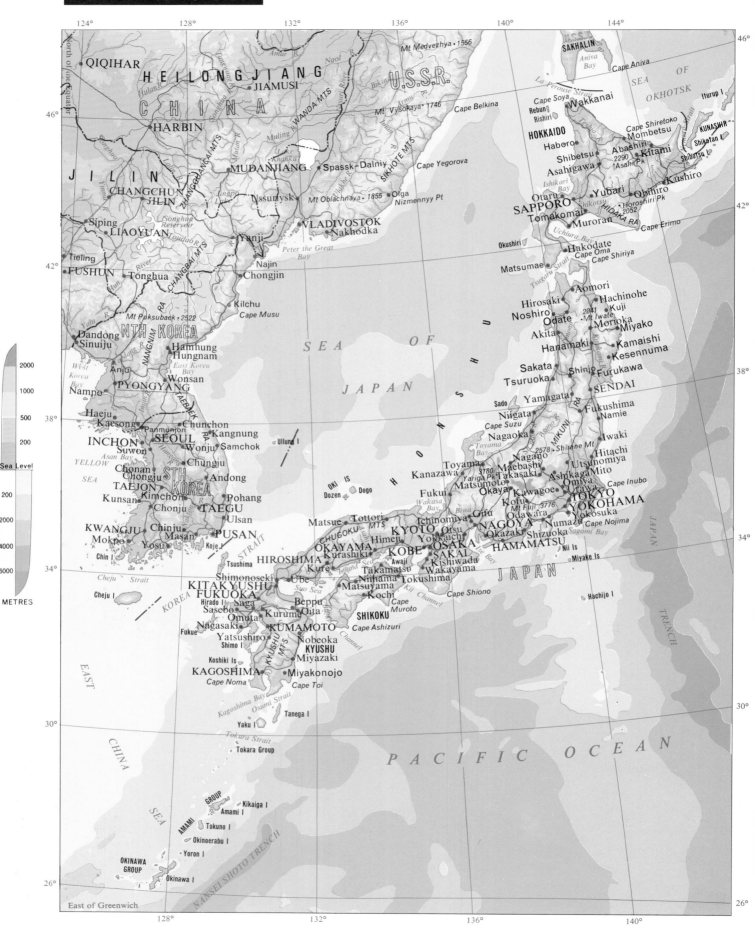

Japan, Korea — physical

QIQIHAR

HEILONGJIANG

JIAMUSI

C H I N A

HARBIN

JILIN

CHANGCHUN
JILIN

Siping
LIAOYUAN

Tieling
FUSHUN

MUDANJIANG

Spassk-Dalniy

Nssuriysk

VLADIVOSTOK
Nakhodka

Tonghua

Yanji

Najin
Chongjin

Dandong
Sinuiju

Kilchu
Cape Musu

NTH KOREA

Hamhung
Hungnam

Anju

Wonsan

Nampo

PYONGYANG

Haeju

Kaesong

Chunchon

Kangnung

SEOUL

Panmunjon

Wonju

Samchok

INCHON

Suwon

Chungju

Chonan

Chongju

Andong

STH
KOREA

TAEJON

Kunsan

Kimchon

Chonju

TAEGU

Pohang

Ulsan

KWANGJU

Chinju

Mokpo

Masan

Yosu

PUSAN

Koje I

U.S.S.R.

Mt Medvezhya • 1556

SAKHALIN

Aniva
Bay

Cape Aniva

SEA

OF

OKHOTSK

Cape Soya

Wakkanai

HOKKAIDO

Haboro

Shibetsu

Asahigawa

Mombetsu

Kitami

Shibotsu

KUNASHR

Shikotan

Asahi Pk
2290

Otaru

Yubari

Obihiro

Kushiro

SAPPORO

Tomakomai

Muroran

Okushiri

Matsumae

Hakodate

Cape Oma

Cape Shiriya

Aomori

Hirosaki

Hachinohe

Kuji

Noshiro

Odate

Mt Iwate • 2041

Morioka

Miyako

Akita

Hanamaki

Kamaishi

Sakata

Shinjo

Kesennuma

Tsuruoka

Furukawa

Yamagata

SENDAI

Sado

Niigata

Fukushima

Cape Suzu

Namie

Nagaoka

Iwaki

Toyama

Nagano

Shirane Mt • 2578

Hitachi

Kanazawa

Maebashi

Utsunomiya

Fukui

Matsumoto

Takasaki

Ashikaga

Mito

Yariga Pk • 3180

Okaya

Kawagoe

Omiya

Cape Inubo

Kofu

Urawa

Ichinomiya

Gifu

Mt Fuji • 3776

TOKYO

Matsue

Tottori

NAGOYA

YOKOHAMA

CHUGOKU MTS

Otsu

Okazaki

Numazu

Yokosuka

KYOTO

Yokkaichi

Shizuoka

Cape Nojima

OKAYAMA

Himeji

OSAKA

HAMAMATSU

Odawara

Sagami Bay

Kurashiki

KOBE

SAKAI

Miyake Is

HIROSHIMA

Kure

Awaji

Kishiwada

Nii Is

Ube

Takamatsu

Wakayama

Shimonoseki

Niihama

Tokushima

KITAKYUSHU

Matsuyama

Kochi

Hachijo I

FUKUOKA

Beppu

Cape Muroto

Hirado I

Saga

Kurume

Oita

Sasebo

Omuta

SHIKOKU

Nagasaki

KUMAMOTO

Cape Ashizuri

Fukue

Yatsushiro

Nobeoka

Shimo I

Miyazaki

Koshiki Is

KYUSHU

KAGOSHIMA

Miyakonojo

Cape Noma

Cape Toi

Kagoshima Bay

Osumi Strait

Tanega I

Yaku I

Tokara Strait

Tokara Group

P A C I F I C   O C E A N

EAST

CHINA

SEA

GROUP

Kikaiga I

Amami

AMAMI

Tokuno I

Okinoerabu I

Yoron I

OKINAWA
GROUP

Okinawa I

NANSEI SHOTO TRENCH

Cheju I

Cheju Strait

YELLOW

SEA

Chin I

Tsushima

KOREA STRAIT

JAPAN TRENCH

S E A   O F

J A P A N

OKI IS

Dozen

Dogo

H O N S H U

Wakasa Bay

Biwa

Inland Sea

Suo Sea

Kii Channel

Cape Shiono

JAPAN

Ullung I

2000
1000
500
200

Sea Level

200
2000
4000
6000

METRES

East of Greenwich

Scale 1:10 100 000 ; 1 centimetre to 101 kilometres
Miller Oblated Stereographic Projection
Heights and depths in metres

KILOMETRES

100 50 0   100   200   300   400

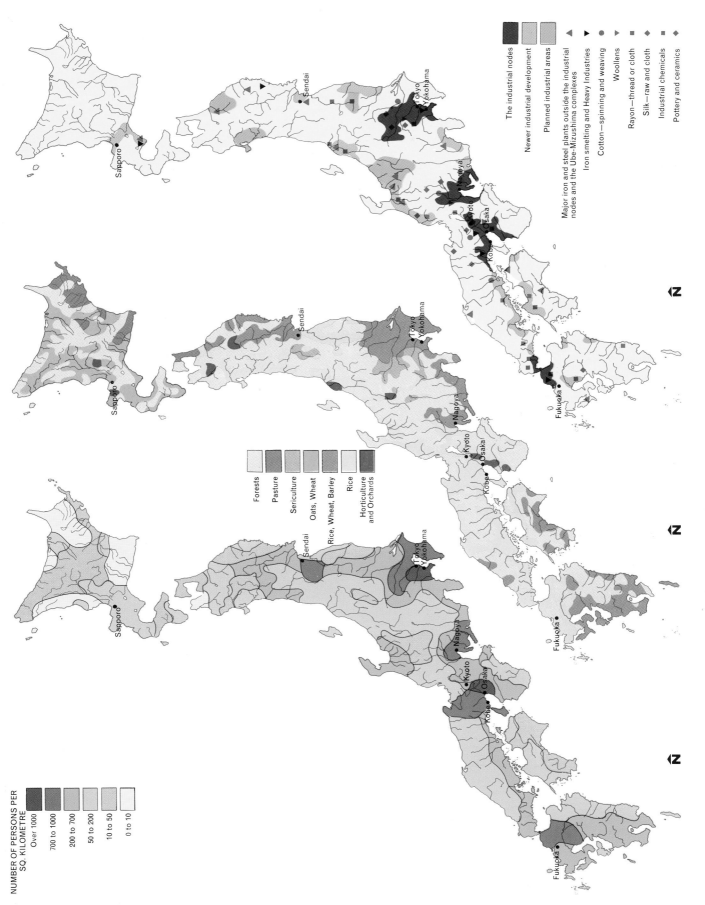

Japan

NUMBER OF PERSONS PER SQ. KILOMETRE

- Over 1000
- 700 to 1000
- 200 to 700
- 50 to 200
- 10 to 50
- 0 to 10

- Forests
- Pasture
- Sericulture
- Oats, Wheat
- Rice, Wheat, Barley
- Rice
- Horticulture and Orchards

- The industrial nodes
- Newer industrial development
- Planned industrial areas

- Major iron and steel plants outside the industrial nodes and the Ube-Mizushima complexes
- Iron smelting and Heavy Industries
- Cotton—spinning and weaving
- Woollens
- Rayon—thread or cloth
- Silk—raw and cloth
- Industrial chemicals
- Pottery and ceramics

Sapporo

Sendai

Tokyo
Yokohama

Nagoya

Kyoto
Osaka
Kobe

Fukuoka

KILOMETRES
100 50 0 100 200 300 400

Scale 1:10 000 000 ; 1 centimetre to 100 kilometres
Miller Oblated Stereographic Projection

A19

Settlement in South-east Asia

BAY OF BENGAL

MYANMAR (BURMA)

Chiang Rai
Chiang Mai
Luang Prabang
Nanning
Pingxiang
CHINA
Foshan
GUANGZHOU

Henzada
Lampang
Nam Dinh
Hanoi
Haiphong
Beihai
Zhanjiang
Victoria
HONG KONG (U.K.)

Bassein
Pegu
YANGON
LAOS

Cape Negrais
Moulmein
Phitsanulok
Udon Thani
Vientiane
Vinh
Xom Hang
Haikou
HAINAN

Mouths of the Irrawaddy
Gulf of Martaban
Ye
Uthai Thani
THAILAND
Phetchabun
Sakon
Savannakhet
Tchepone
Hue
Yacheng
Ya Xian

Tavoy
Nakhon Thai
Khon Kaen
Chaiyaphum

North Andaman
ANDAMAN
Middle Andaman
ISLANDS
Nakhon Pathom
Lop Buri
Nakhon Ratchasima
Surin
Pakse
Da Nang
Quang Ngai
Dak Robong

South Andaman
ANDAMAN SEA
Samut Songkhram
Ayutthaya
Ban Bok
Koulen
VIETNAM
Qui Nhon

BANGKOK
Chon Buri
Tuy Hoa

10°N
Phet Buri
Rayong
Battambang
Moung
CAMBODIA
Sre Sbau
Kampong Chhnang

Mergui
Mergui
Pouthisat
Kampong Cham
Da Lat
Nha Trang
Camranh

Archipelago
GULF
Phnom Penh
Banam
Phan Rang

OF
Kampot
Long Xuyen
My Tho
HO CHI MINH CITY
PHILIPP

Surat Thani
THAILAND
Rach Gia
Can Tho
Vung Tau
SOUTH
CHINA

Phangan
Point Baibung
Quan Long
Mouths of the Mekong
SEA

Phuket Island
Phuket
Nakhon Si Thammarat

Trang
Phatthalung

Sabang
Banda Aceh
Sigli
Songkhla
Pattani
Puerto Princesa
Quezon
Eran
Tay

Meulaboh
Pantonlabu
Hat Yai
Narathiwat
Langkawi
Kota Baharu
Bonobon

Leuser 3380m
Pangkalansusu
Alor Setar
Georgetown
Pinang
Butterworth
Balabac
Balabac
Balabac Strait

Sibigo
Simeulue
Ipoh
Kuala Terengganu
Sikuati

MEDAN
WEST MALAYSIA
Dungun
M A L A Y S I A
Kota Kinabalu
Kinabalu 4101m
Sand

Pematangsiantar
Kuala Selangor
Kelang
Kuantan
Mempakul

Banjak
Barus
Lahewa
Kelang
Kuala Lumpur
Bandar Seri Begawan
BRUNEI
Pensiangan

Sibolga
Nias
Bagansiapiapi
Langgapajung
Seremban
Melaka
Tioman
Natuna
Anambas
Miri
EAST MALAYSIA
Sibuku
Taw

Telukdalam
Sebanga
Meskum
Kudap
South Natuna
Balingian
Bintulu
Bario
Tarakan
Lemutan
Longleju
Salimbatu
Tanjungselo

Equator 0°
Pini
Airbangis
Sawang
Johor Baharu
SINGAPORE
Cape Datu
Paloh
Oya
Belaga
Napaku

Batu Islands
Pakanbaru
Burung
Bintan
SINGAPORE
Lundu
Sibu
Kok
BORNEO

Sigep
Singkarak
Bukittinggi
Sawahlunto
Lirik
Lingga
Singkep
Kuching
Sungaibatu
Kubumesaai

Siberut
Muarasiberut
Padang
SUMATRA
Sungaisalak
Mempawah
Sungaibatu
Mualang
Menjapa 2000m
Batukelau
Muaramawai
Sepasu

Sigoisooinan
Sipura
Kambang
Kerinci 3805m
Sungaipenuh
Simpang
Telukmajebi
Pontianak
Sintang
Bontang

North Pagai
Bake
South Pagai
Islands
Ipuh
Mukomuko
Sarolangun
Kampa
Bangka
Kotabaharu
Raja 2278m
Rangantemiang
Kualakurun
Samarinda

Ketaun
Penuguan
Sungsang
Muntok
Karimata
Nangatajap
Panahan
Bakumpai
Timpah
Daju
Balikpap

Bengkulu
Tais
Lahat
Pangkalpinang
Koba
Kualapesaguan
KALIMANTAN
Palangkaraya
Mapan
Mehakit

Palembang
Petaling
Batubetumpang
Toboali
Belitung
Kendawangan
Bangkal
Satui
Laut

Pagerdewa
Talangbatu
Cape Puting
Banjarmasin

Manna
Martapura
Cape Selatan
Makasar

Enggano
Kotabumi
Gunungsugih
JAVA
Bawean
SEA
Sump

Kru
Belimbing
Kalianda
Tanjungkarang
Ujung P

INDIAN OCEAN
Serang
Bogor
JAKARTA
Cirebon
Tegal
Pekalongan
SEMARANG
Kudus
Surakarta
Tuban
Sumenep

Sukabumi
BANDUNG
Sindangbarang
Cipatuja
Magelang
Surakarta
Blitar
Malang
Singaraja
Denpasar
Mataram

10°S
Jogyakarta
JAVA
Madiun
Kediri
Probolinggo
Puger
Situbondo
Bali
Lombok

A20

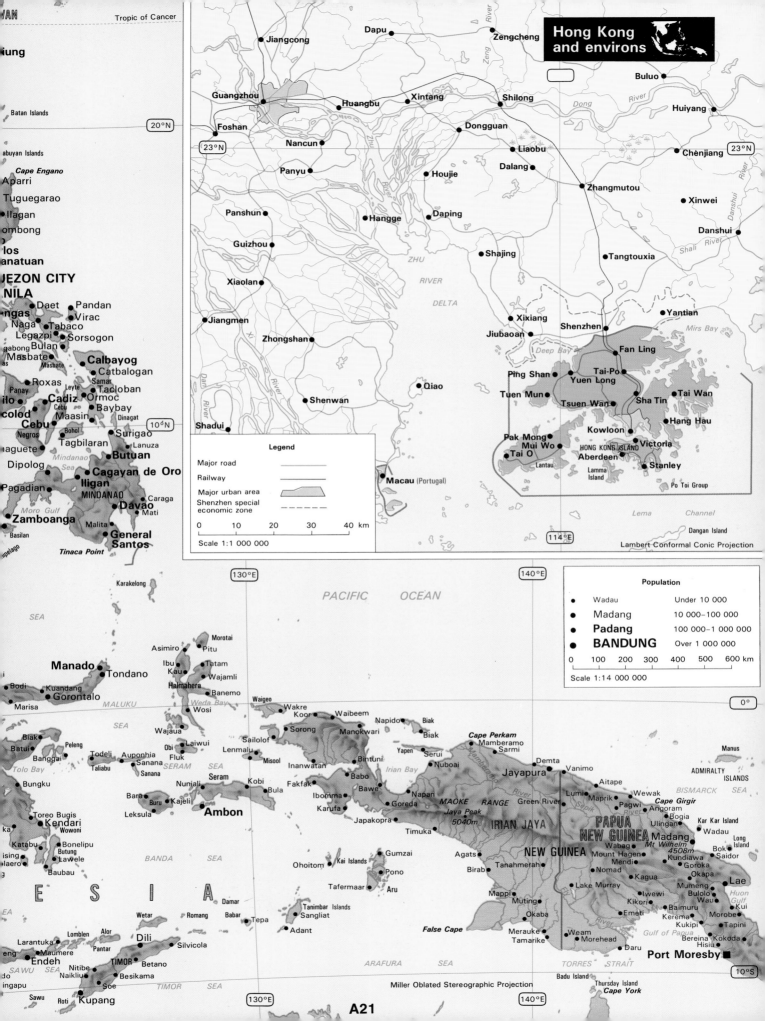

Jiangcong
Dapu
Zengcheng

Buluo

Hong Kong and environs

Guangzhou
Huangbu
Xintang
Shilong

Huiyang

Foshan

Dongguan
Chènjiang

23°N
Nancun
Liaobu
23°N

Panyu
Houjie
Dalang
Zhangmutou
Xinwei

Panshun
Hangge
Daping

Danshui

Guizhou
Shajing
Tangtouxia

Xiaolan

Yantian

Jiangmen
Xixiang
Shenzhen
Fan Ling

Zhongshan
Jiubaoan
Ping Shan
Tai-Po

Yuen Long
Tai Wan

Qiao
Tuen Mun
Tsuen Wan
Sha Tin

Shenwan
Hang Hau

Shadui
Pak Mong
Kowloon
Victoria

Mui Wo
HONG KONG ISLAND

Tai O
Aberdeen
Stanley

Macau (Portugal)
Lantau
Lamma Island
Po Toi Group

Legend

Major road
Railway
Major urban area
Shenzhen special economic zone

0   10   20   30   40 km

Scale 1:1 000 000

114°E

Dangan Island

Lema Channel

Lambert Conformal Conic Projection

Batan Islands

20°N

Cape Engano
Aparri
Tuguegarao
Ilagan
ombong
los
anatuan
QUEZON CITY
NILA
ngas
Daet
Pandan
Naga
Virac
Legazpi
Tabaco
Sorsogon
gabong
Bulan
Masbate
Masbate
Calbayog
Catbalogan
Roxas
Samar
Tacloban
Panay
Leyte
Ormoc
ilo
Cadiz
Cebu
Baybay
colod
Maasin
Dinagat
Cebu
Bohol
colod
Negros
Surigao
aguete
Tagbilaran
Lanuza
Dipolog
Butuan
Pagadian
Iligan
Cagayan de Oro
MINDANAO
Caraga
Zamboanga
Davao
Mati
Moro Gulf
Malita
Basilan
General Santos
Tinaca Point
pelago

10°N

Mindanao Sea

Population

•	Wadau	Under 10 000
•	Madang	10 000–100 000
•	**Padang**	100 000–1 000 000
●	**BANDUNG**	Over 1 000 000

0   100   200   300   400   500   600 km

Scale 1:14 000 000

0°

130°E
140°E

PACIFIC   OCEAN

Karakelong

SEA

Morotai
Asimiro
Pitu
Ibu
Tatam
Kau
Wajamli
Manado
Halmahera
Banemo
Tondano
Bodi
Kuandang
Waigeo
Wakre
Koor
Waibeem
Napido
Biak
Gorontalo
MALUKU
Weda Bay
Wosi
Sorong
Manokwari
Biak
Marisa
Wajaua
Sailolof
Napido

Biak
Laiwui
Obi
Lenmalu
Yapen
Serui
Cape Perkam
Mamberamo
Sarmi
Batui
Peleng
Fluk
Misool
Nuboai
Demta
Vanimo
Manus
Banggai
Todeli
Auponhia
Inanwatan
Irian Bay
Jayapura
ADMIRALTY ISLANDS
Taliabu
Sanana
Babo
Aitape
Tolo Bay
Bungku
Seram
Kobi
Bula
Fakfak
Bawe
Lumi
Maprik
Wewak
Cape Girgir
BISMARCK SEA
Bara
Nunjali
Ibomma
Napan
MAOKE   RANGE
Green River
Pagwi
Angoram
SERAM   SEA
Buru
Kajeli
Goreda
Jaya Peak 5040m
Wabag
Bogia
Kar Kar Island
Leksula
Ambon
Karufa
Japakopra
IRIAN JAYA
Mount Hagen
Ulingan
Wadau
Toreo Bugis
Wowoni
Timuka
NEW GUINEA
Mendi
Madang
Long Island
Kendari
PAPUA NEW GUINEA
Mt Wilhelm 4508m
Katabu
Boneliptu
Gumzai
Agats
Tanahmerah
Nomad
Kundiawa
Bok
Saidor
ising
Butung
Lawele
Kai Islands
Pono
Birab
Lake Murray
Kagua
Goroka
Okapa
laero
Baubau
Aru
Mappi
Muting
Iwewi
Mumeng
Lae
ESIA
Ohoitom
Tafermaar
Okaba
Kikori
Bulolo
Wau
Morobe
Tanimbar Islands
Sangliat
Merauke
Kerema
Kukipi
Tapini
Damar
Adant
False Cape
Weam
Morehead
Emeti
Bereina
Hisiu
Wetar
Romang
Babar
Tepa
Muting
Daru
Kokoda
Lomblen
Alor
Tamarike
Port Moresby
Larantuka
Pantar
Silvicola
Maumere
Dili
ARAFURA   SEA
eng
Nitibe
Betano
TORRES   STRAIT
Endeh
Naikliu
Besikama
Badu Island
Thursday Island
ingapu
Soe
TIMOR
SAWU SEA
Cape York
Sawu
Roti
Kupang

Banda Sea

BANDA   SEA

TIMOR   SEA

130°E
140°E
10°S

Miller Oblated Stereographic Projection

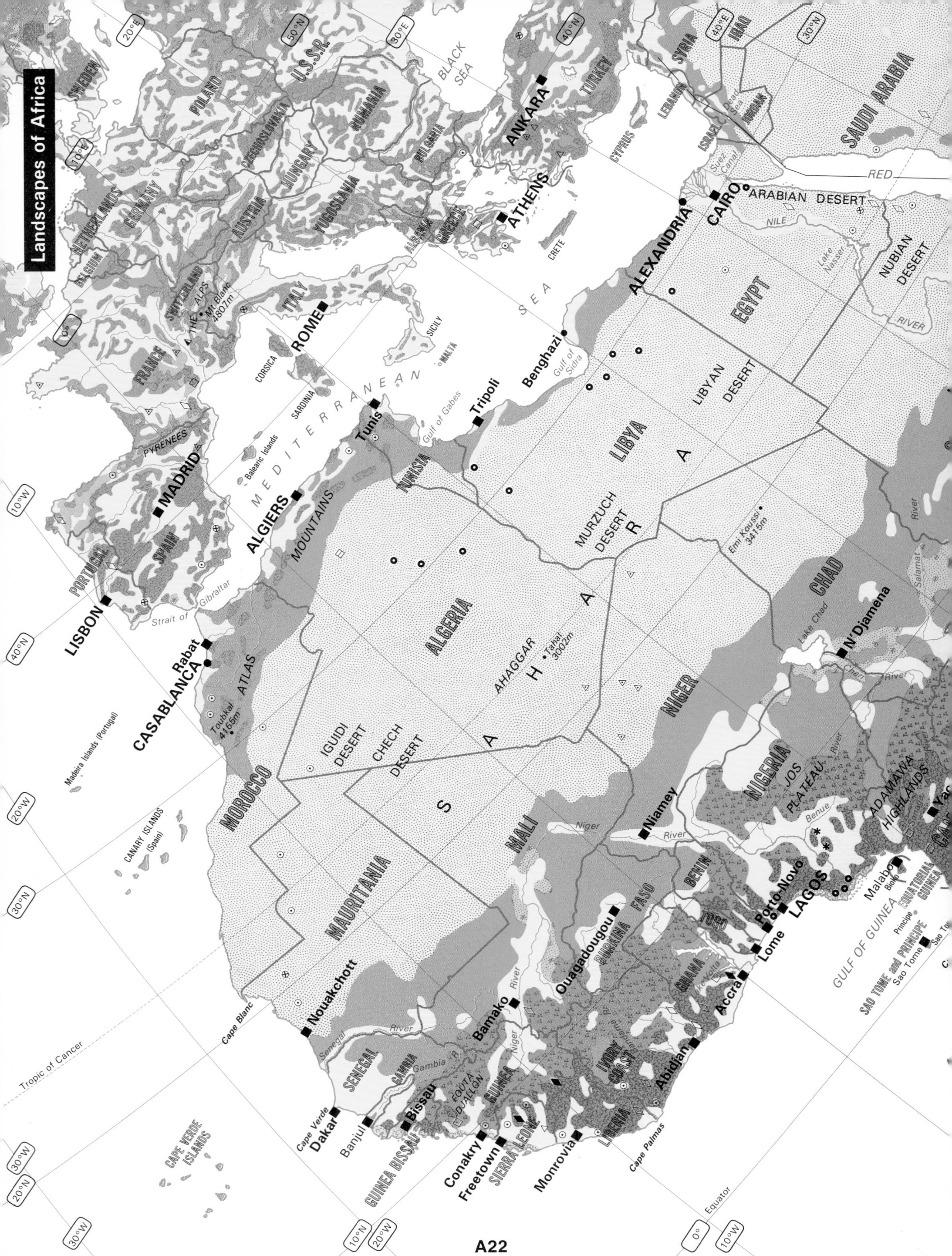

SWEDEN

NETHERLANDS
BELGIUM
GERMANY
POLAND
U.S.S.R.
SWITZERLAND
FRANCE
AUSTRIA
HUNGARY
CZECHOSLOVAKIA
YUGOSLAVIA
ROMANIA
BULGARIA
ITALY
ALBANIA
GREECE
ATHENS
TURKEY
ANKARA

BLACK SEA

SYRIA
IRAQ
LEBANON
CYPRUS
ISRAEL
JORDAN
Dead Sea
SAUDI ARABIA

THE ALPS
Mt. Blanc
4807m
CORSICA
SARDINIA
SICILY
MALTA
CRETE

SEA

ALEXANDRIA
CAIRO
ARABIAN DESERT
NILE
Lake Nasser
RED
NUBIAN DESERT

ROME
MADRID
SPAIN
PORTUGAL
LISBON
Balearic Islands
MEDITERRANEAN
Gibraltar
Strait of
PYRENEES

Tunis
TUNISIA
ALGIERS
MOUNTAINS
ATLAS
Rabat
CASABLANCA
Toubkal
4165m

Tripoli
Benghazi
Gulf of Sidra
Gulf of Gabes

LIBYA
MURZUCH DESERT
LIBYAN DESERT
EGYPT

S A H A R A

ALGERIA
AHAGGAR
Tahat
3002m
Emi Koussi
3415m

Madeira Islands (Portugal)

MOROCCO
IGUIDI DESERT
CHECH DESERT

CANARY ISLANDS
(Spain)

MAURITANIA

Nouakchott
Cape Blanc

MALI
Niger
S
Niamey
NIGER
CHAD
N'Djamena
Lake Chad
Chari
River
Salamat

NIGERIA
JOS PLATEAU
Benue
ADAMAWA HIGHLANDS
Yaoundé

CAPE VERDE ISLANDS

Cape Verde
Dakar
SENEGAL
Senegal River
Banjul
GAMBIA
Gambia R.
GUINEA BISSAU
Bissau
FOUTA DJALLON
GUINEA
Conakry
Freetown
SIERRA LEONE
Monrovia
LIBERIA
Cape Palmas

Bamako
Niger River
Ouagadougou
BURKINA FASO
Black Volta
IVORY COAST
Abidjan
GHANA
Accra
TOGO
Lome
BENIN
Porto-Novo
LAGOS
Malabo
Bioko
EQUATORIAL GUINEA
SAO TOME and PRINCIPE
Principe
Sao Tome
Sao Tomé
GULF OF GUINEA

Tropic of Cancer
Equator

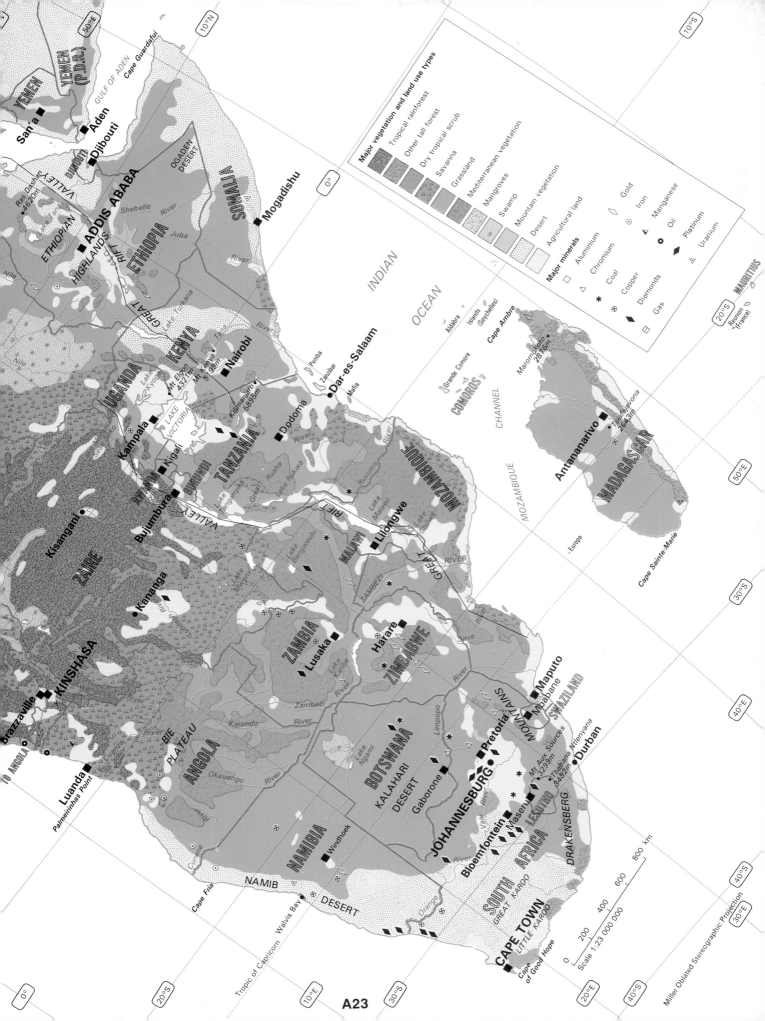

Landscapes of Europe

Major vegetation and land use types
- Tall forest
- Grassland
- Mediterranean vegetation
- Swamp
- Mountain vegetation
- Tundra
- Desert
- Agricultural land

Major minerals

Symbol	Mineral	Symbol	Mineral
□	Aluminium	▲	Lead
△	Chromium	◮	Manganese
✳	Coal	☆	Mercury
⊗	Copper	⊙	Oil
⊟	Gas	⬦	Tungsten
◇	Gold	△	Uranium
⊙	Iron	⊞	Zinc

Scale 1:15 000 000

0   200   400   600 km

BARENTS SEA

Murmansk
KOLA PENINSULA
KANIN PENINSULA
Kolguyev
WHITE SEA
Gulf of Onega
Arkhangelsk
North Dvina River
Lake Onega
Lake Beloye
Lake Ladoga
LENINGRAD
Lake Chudskoye
Andropov Reservoir
Volga River
MOSCOW
KAZAN
GORKI
Kama River
PERM
SVERDLOVSK
CHELYABINSK
UFA
URAL MOUNTAINS
Ob River
Irtysh River
KUIBYSHEV
UNION OF SOVIET SOCIALIST REPUBLICS
Voronezh
Don River
KHARKOV
KIEV
Volga
Volgograd
Ural River
ARAL SEA
Syr River
KYZYL DESERT
Dnieper River
DONETSK
Rostov
ODESSA
SEA OF AZOV
BUCHAREST
CAUCASUS
Mt Elbrus 5642m
MOUNTAINS
TBILISI
CASPIAN SEA
BAKU
BLACK SEA
Samsun
YEREVAN
Lake Sevan
Mt Ararat 5165m
ISTANBUL
Erzurum
Tabriz
ANKARA
Lake Van
Lake Urmia
TURKEY
IZMIR
KONYA
Lake Tuz
ADANA
IRAN
Mosul
Rhodes
Nicosia
CYPRUS
Aleppo
SYRIA
Euphrates River
Tigris River
IRAQ
A25
BAGHDAD

40°E 50°E 60°E 70°E 80°E
70°N 60°N 50°N 40°N
30°E 40°E 50°E

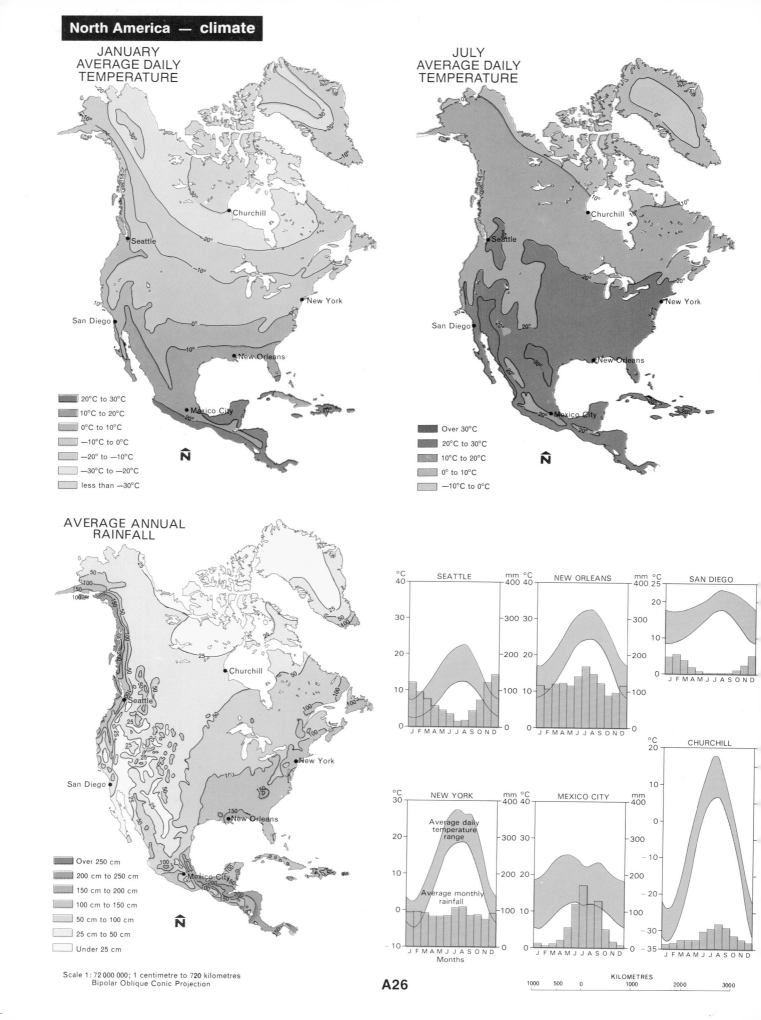

North America — climate

JANUARY AVERAGE DAILY TEMPERATURE

20°C to 30°C
10°C to 20°C
0°C to 10°C
−10°C to 0°C
−20° to −10°C
−30°C to −20°C
less than −30°C

Churchill
Seattle
New York
San Diego
New Orleans
Mexico City

N̂

JULY AVERAGE DAILY TEMPERATURE

Over 30°C
20°C to 30°C
10°C to 20°C
0° to 10°C
−10°C to 0°C

Churchill
Seattle
New York
San Diego
New Orleans
Mexico City

N̂

AVERAGE ANNUAL RAINFALL

Over 250 cm
200 cm to 250 cm
150 cm to 200 cm
100 cm to 150 cm
50 cm to 100 cm
25 cm to 50 cm
Under 25 cm

Churchill
Seattle
New York
San Diego
New Orleans
Mexico City

N̂

Scale 1 : 72 000 000; 1 centimetre to 720 kilometres
Bipolar Oblique Conic Projection

SEATTLE
NEW ORLEANS
SAN DIEGO
NEW YORK
Average daily temperature range
Average monthly rainfall
Months
MEXICO CITY
CHURCHILL

A26

KILOMETRES
1000 500 0 1000 2000 3000

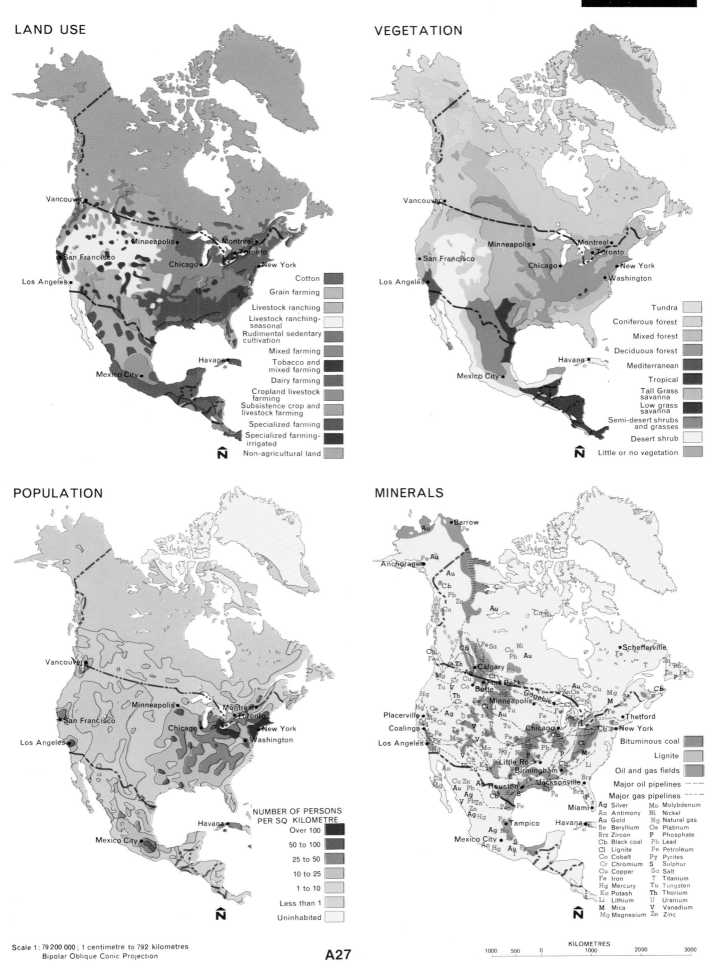

LAND USE

- Cotton
- Grain farming
- Livestock ranching
- Livestock ranching-seasonal
- Rudimental sedentary cultivation
- Mixed farming
- Tobacco and mixed farming
- Dairy farming
- Cropland livestock farming
- Subsistence crop and livestock farming
- Specialized farming
- Specialized farming-irrigated
- Non-agricultural land

Vancouver, Minneapolis, Montreal, Toronto, San Francisco, Chicago, New York, Los Angeles, Havana, Mexico City

VEGETATION

- Tundra
- Coniferous forest
- Mixed forest
- Deciduous forest
- Mediterranean
- Tropical
- Tall Grass savanna
- Low grass savanna
- Semi-desert shrubs and grasses
- Desert shrub
- Little or no vegetation

Vancouver, Minneapolis, Montreal, Toronto, San Francisco, Chicago, New York, Washington, Los Angeles, Havana, Mexico City

POPULATION

NUMBER OF PERSONS PER SQ KILOMETRE
- Over 100
- 50 to 100
- 25 to 50
- 10 to 25
- 1 to 10
- Less than 1
- Uninhabited

Vancouver, Minneapolis, Montreal, Toronto, San Francisco, Chicago, New York, Washington, Los Angeles, Havana, Mexico City

MINERALS

- Bituminous coal
- Lignite
- Oil and gas fields
- Major oil pipelines
- Major gas pipelines

Ag	Silver	Mo	Molybdenum
An	Antimony	Ni	Nickel
Au	Gold	Ng	Natural gas
Be	Beryllium	Os	Platinum
Brz	Zircon	P	Phosphate
Cb	Black coal	Pb	Lead
Cl	Lignite	Pe	Petroleum
Co	Cobalt	Py	Pyrites
Cr	Chromium	S	Sulphur
Cu	Copper	Sa	Salt
Fe	Iron	T	Titanium
Hg	Mercury	Tu	Tungsten
Ka	Potash	Th	Thorium
Li	Lithium	U	Uranium
M	Mica	V	Vanadium
Mg	Magnesium	Zn	Zinc

Barrow, Pe, Anchorage, Calgary, Fort Peck, Butte, Gogebic, Thetford, Placerville, Coalinga, Minneapolis, Chicago, New York, Los Angeles, Little Rock, Birmingham, Scheffervville, Houston, Jacksonville, Miami, Tampico, Havana, Mexico City

KILOMETRES
1000 500 0 1000 2000 3000

Scale 1:11 000 000; 1 centimetre to 110 kilometres
Bipolar Oblique Conic Projection
Heights and depths in metres

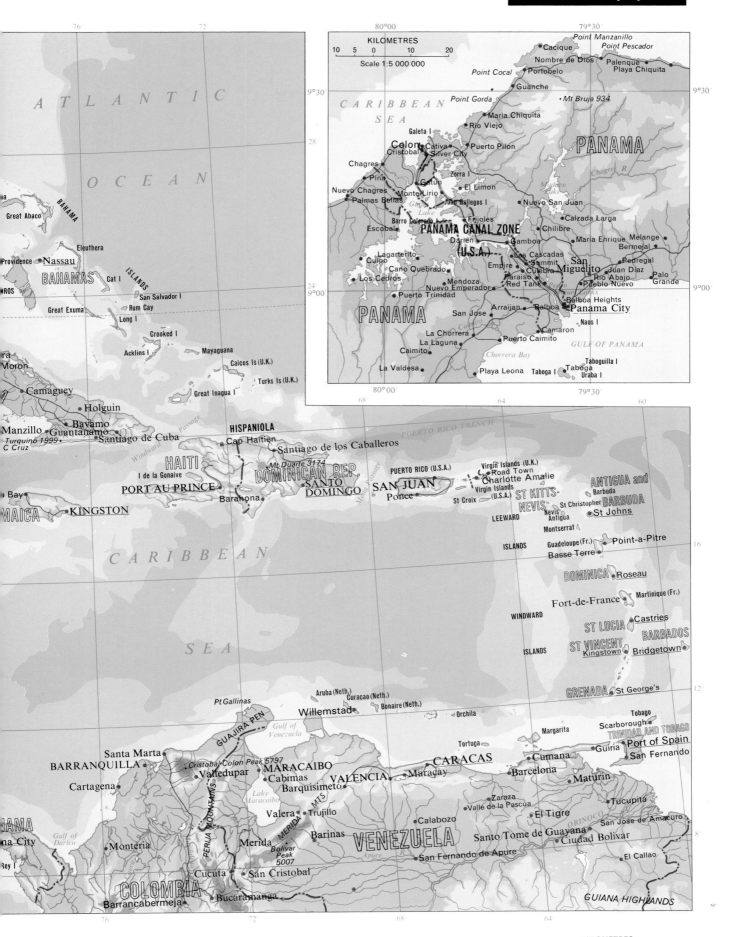

ATLANTIC
OCEAN

Great Abaco

Providence Nassau
Eleuthera
BAHAMAS Cat I
ISLANDS
San Salvador I
Great Exuma Rum Cay
Long I
Crooked I
Acklins I Mayaguana
Caicos Is (U.K.)
Great Inagua
Turks Is (U.K.)

ra
Moron

Camaguey
Holguin
Bayamo
Manzillo Guantanamo
C Cruz Turquino 1999 Santiago de Cuba

HISPANIOLA
Cap Haitien
Santiago de los Caballeros
Windward
Passage Mt Duarte 3174
HAITI DOMINICAN REP.
I de la Gonaive
PORT AU PRINCE SANTO
Bay Barahona DOMINGO SAN JUAN
Ponce
MAICA St Croix
KINGSTON

CARIBBEAN

SEA

Pt Gallinas
Aruba (Neth.)
Curacao (Neth.)
GUAJIRA PEN Bonaire (Neth.)
Willemstad
Gulf of
Venezuela
Santa Marta
BARRANQUILLA Cristobal Colon Peak 5797
Valledupar
Cartagena Cabimas VALENCIA
MARACAIBO
Barquisimeto
Lake MTS
AMA Maracaibo
na City Valera Trujillo
Gulf of MERIDA
Darien Barinas VENEZUELA
Monteria Merida
Bolivar
Peak
Rey I 5007 Apure
COLOMBIA Cucuta San Cristobal
Barrancabermeja Bucaramanga

PUERTO RICO TRENCH

PUERTO RICO (U.S.A.)
Virgin Islands (U.K.)
Road Town
Charlotte Amalie
Virgin Islands
(U.S.A.)
ST KITTS-
NEVIS St Christopher
LEEWARD Nevis
Antigua
Montserrat
ISLANDS Guadeloupe (Fr.)
Basse Terre
DOMINICA Roseau
Fort-de-France Martinique (Fr.)
WINDWARD
ST LUCIA Castries
ISLANDS ST VINCENT BARBADOS
Kingstown Bridgetown

GRENADA St George's

ANTIGUA and
Barbuda
BARBUDA
St Johns

Point-a-Pitre

Tobago
Scarborough TRINIDAD AND TOBAGO
Margarita Port of Spain
Tortuga Guiria San Fernando
CARACAS Cumana
Maracay Barcelona
Maturin
Zaraza
Valle de la Pascua Tucupita
Calabozo El Tigre
Santo Tome de Guayana San Jose de Amacuro
San Fernando de Apure Ciudad Bolivar
ORINOCO
El Callao
GUIANA HIGHLANDS

KILOMETRES
80°00 79°30
Point Manzanillo
Cacique Point Pescador
Nombre de Dios Palenque
Point Cocal Portobelo Playa Chiquita
Guanche
CARIBBEAN Point Gorda • Mt Bruja 934
SEA Maria Chiquita
Rio Viejo
Galeta I PANAMA
28 Colon Cativa Puerto Pilon
Cristobal Silver City
Chagres Zorra
Pina Gatun El Limon
Nuevo Chagres Monte Lirio
Palmas Bellas Juan Gallegos I Nuevo San Juan
Calzada Larga
Barro Colorado Frijoles
Escobal PANAMA CANAL ZONE Chilibre
24 Darien (U.S.A.) Gamboa Maria Enrique Melange
Lagarterito Las Cascadas Bermejal
Cuipo Empire Summit Pedregal
Cano Quebrado Culebra San Juan Diaz
Los Cedros Paraiso Miguelito Rio Abajo Palo
Mendoza Red Tank Rio Abajo Grande
9°00 Nuevo Emperador Pueblo Nuevo
Puerto Trinidad Balboa Heights
PANAMA Arraijan Balboa Panama City
San Jose Naos I
Camaron
La Chorrera Puerto Caimito
La Laguna GULF OF PANAMA
Caimito
Chorrera Bay Taboguilla I
La Valdesa Taboga
Playa Leona Taboga I Uraba I

KILOMETRES
10 5 0 10 20
Scale 1:5 000 000

KILOMETRES
100 50 0 100 200 300 400

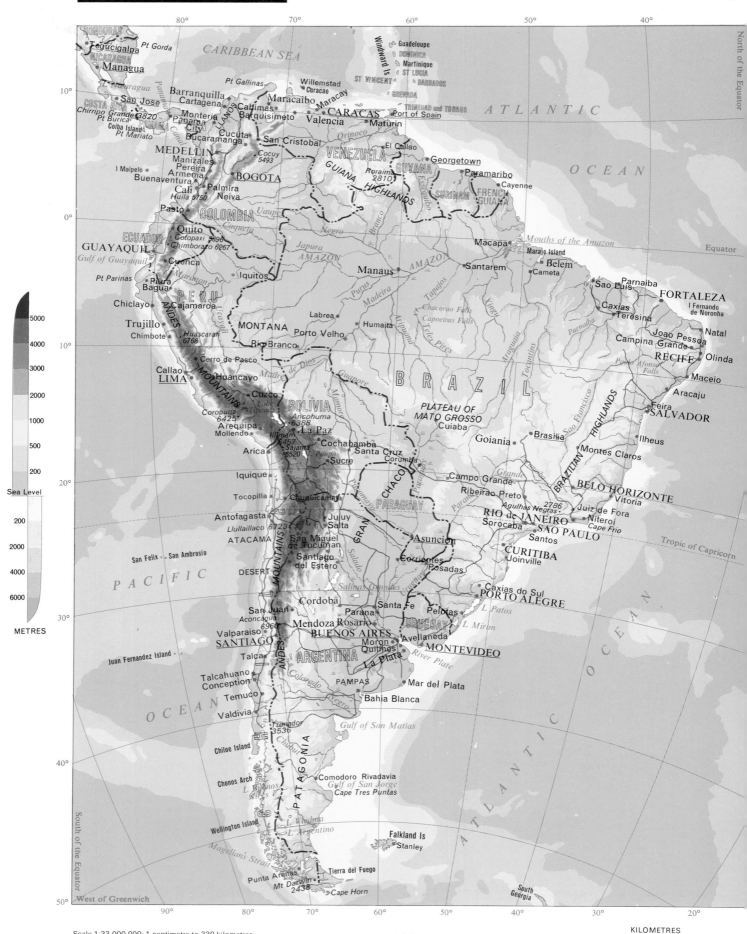

Scale 1:33 000 000; 1 centimetre to 330 kilometres
Bipolar Oblique Conic Projection
Heights and depths in metres

A30

KILOMETRES
400 200 0 400 800 12

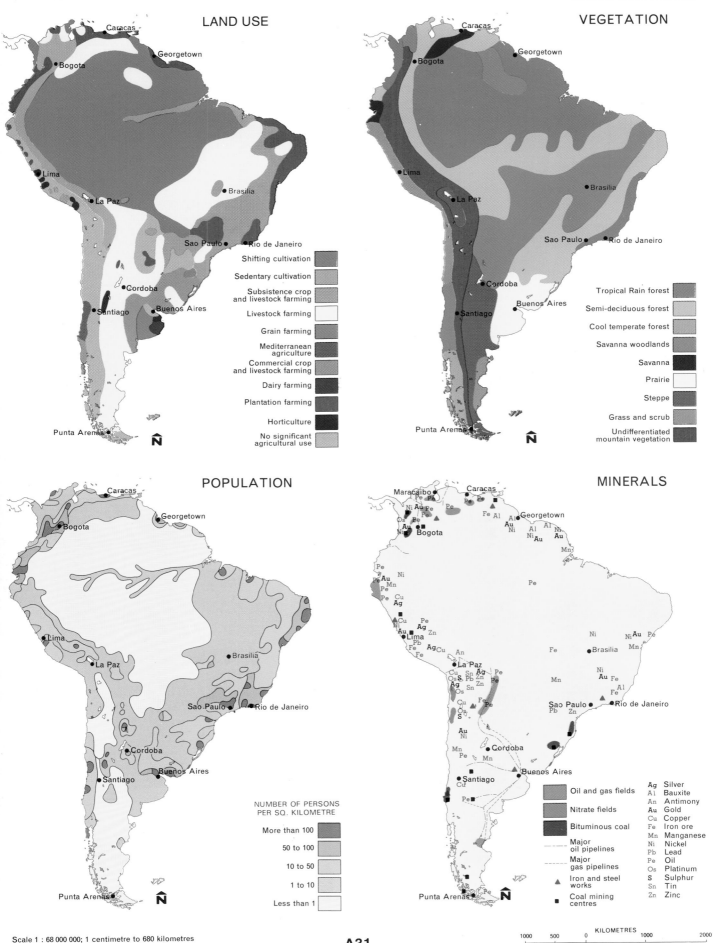

LAND USE

Shifting cultivation
Sedentary cultivation
Subsistence crop and livestock farming
Livestock farming
Grain farming
Mediterranean agriculture
Commercial crop and livestock farming
Dairy farming
Plantation farming
Horticulture
No significant agricultural use

VEGETATION

Tropical Rain forest
Semi-deciduous forest
Cool temperate forest
Savanna woodlands
Savanna
Prairie
Steppe
Grass and scrub
Undifferentiated mountain vegetation

POPULATION

NUMBER OF PERSONS PER SQ. KILOMETRE

More than 100
50 to 100
10 to 50
1 to 10
Less than 1

MINERALS

Oil and gas fields
Nitrate fields
Bituminous coal
Major oil pipelines
Major gas pipelines
Iron and steel works
Coal mining centres

Ag Silver
Al Bauxite
An Antimony
Au Gold
Cu Copper
Fe Iron ore
Mn Manganese
Ni Nickel
Pb Lead
Pe Oil
Os Platinum
S Sulphur
Sn Tin
Zn Zinc

Scale 1 : 68 000 000; 1 centimetre to 680 kilometres
Miller Oblated Stereographic Projection

A31

KILOMETRES
1000   500   0   1000   2000

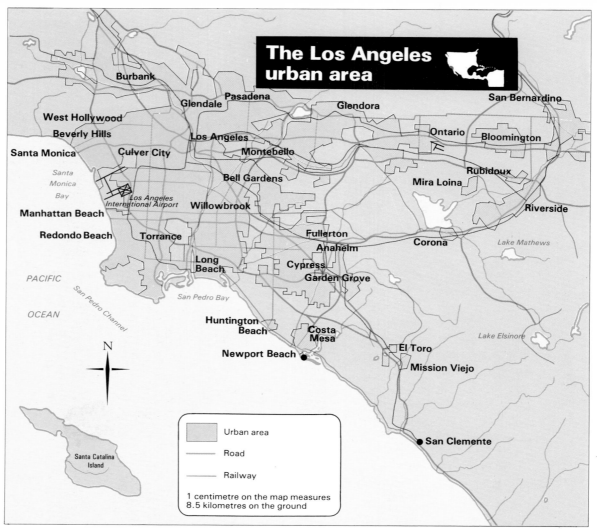

The Los Angeles urban area

Burbank
Glendale
Pasadena
Glendora
San Bernardino
West Hollywood
Ontario
Bloomington
Beverly Hills
Los Angeles
Santa Monica
Culver City
Montebello
Rubidoux
Santa Monica Bay
Bell Gardens
Mira Loina
Manhattan Beach
Los Angeles International Airport
Willowbrook
Riverside
Redondo Beach
Torrance
Fullerton
Corona
Lake Mathews
Anaheim
PACIFIC
Long Beach
Cypress
OCEAN
Garden Grove
San Pedro Channel
San Pedro Bay
Lake Elsinore
Huntington Beach
Costa Mesa
El Toro
Newport Beach
Mission Viejo
N
San Clemente
Santa Catalina Island

	Urban area
	Road
	Railway

1 centimetre on the map measures
8.5 kilometres on the ground

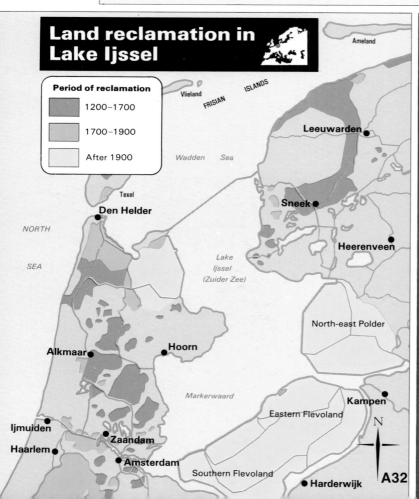

Land reclamation in Lake Ijssel

Ameland

Period of reclamation

	1200–1700
	1700–1900
	After 1900

Vlieland
FRISIAN ISLANDS
Wadden Sea
Leeuwarden
Texel
Sneek
Den Helder
Heerenveen
NORTH
SEA
Lake Ijssel (Zuider Zee)
North-east Polder
Alkmaar
Hoorn
Markerwaard
Kampen
Ijmuiden
Eastern Flevoland
Zaandam
Haarlem
Amsterdam
Southern Flevoland
Harderwijk
N
A32

Landsat scene of Lake Ijssel

1 centimetre on the map and on
the Landsat scene measures
10 kilometres on the ground

Atlas Notes

The following questions and answers, and the "Your Turn" questions, will help you to better understand the maps in the atlas section of this book.

PROJECTIONS

Many maps show the continents and oceans with different shapes and sizes. On some maps, the maps are actually split in places.

Q. *Why do maps of the same area have different shapes and sizes?*

A. The differences are caused when the spherical surface of the globe is forced onto a flat piece of paper. The various ways in which this is accomplished are called **projections**. Some projections are accurate in shape, others in size, and others in direction. No projection can be accurate in all of these respects simultaneously. Map makers choose what they consider to be the most appropriate projection for their needs.

Your Turn

1. a) Find four maps that show the whole world. Each one should have a different projection.
    b) Explain how each differs from the others.
    c) Suggest reasons why each projection was chosen.

RATIO SCALES

Ratio scales look like "1:200 000 000" (see page A1).

Q. *How can I make ratio scales easy to use?*

A. Move the decimal point five places to the left. Thus, "1:200 000 000" can be read "1 cm represents 2 000 km". This is called a written scale.

Your Turn

2. Give a written scale for the maps on pages A6 and A7.

LANDSAT IMAGES

If you look at the Landsat scene on page A15, you will see that the cultivated Nile area appears brownish.

Q. *Why are the colours in the Landsat scenes not always true colours?*

A. Humans see only a limited number of wavelengths. Sensors on satellites record wavelengths outside the normal light spectrum. This is done to give additional types of information.

Your Turn

3. a) What kinds of additional information can you observe in the Landsat image on page A15 as opposed to the map above it?
    b) In what ways is the map of agricultural land use more useful?

DIRECTION ARROWS

North arrows are shown on some maps (see pages A19 and A26), but not on others.

Q. *Is a map always drawn with north at the top?*

A. In the majority of maps, north is somewhere at the top. A problem arises when the projection does not allow lines of longitude to be drawn parallel. In these cases, you can judge the north direction by the direction of the lines of longitude, and the north direction will vary from one part of the map to another. See map on pages A16 and A17.

Your Turn

4. Look at the map on pages A16 and A17 (Landscapes of Asia). If you had to draw a north arrow at each of the following places, would the arrow point up, down, left, or right?
    - London
    - Kabal
    - Beijing
    - St. Lawrence Island (Bering Sea)

5. a) Find a large map in this atlas section where north is far from the top centre of the map.
    b) Sketch or describe the appearance of this map.
    c) Explain why the map has been drawn in this unusual way.

8

Egypt
A RIVER-DEPENDENT REGION

Figure 8.a *Cairo, Egypt's Capital*

Figure 8.b *The Nile River, Aswan*

Figure 8.c *The Sinai Peninsula*

Figure 8.d *Abu Simbel, Tomb of Ramses II*

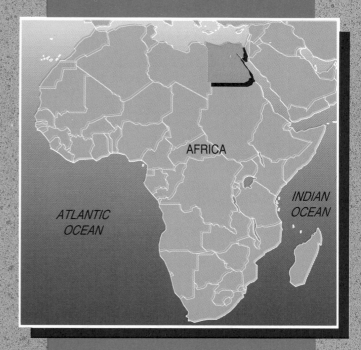

AFRICA

ATLANTIC OCEAN

INDIAN OCEAN

Figure 8.e *The Sahara Desert*

STARTING OUT

1. Describe Egypt's location by comparing it to:
   - major water bodies;
   - nearby countries;
   - your home.
2. Based on what you observe in these photos, comment on the extreme differences in the physical environment of Egypt.
3. List three ways in which you think the Nile River could be used by Egyptians.
4. List five attractions that would draw tourists to Egypt.

LEARNING DESTINATIONS

By the end of this chapter, you should be able to:
- understand what constitutes a river-dependent region;
- describe the physical environment of Egypt;
- appreciate the importance of the Nile River to the economy and way of life in Egypt;
- recognize the tourist potential of this country;
- understand the importance of some key developments taking place in Egypt.

Introduction

A lthough the world's earliest civilizations were widely scattered across Asia and Africa, they had one characteristic in common — they were all located along rivers. The Assyrians in the Tigris-Euphrates Valley, the people of Mohenjo-Daro along the Indus River, the Shang dynasty of the Hwang Ho Valley, and the Egyptian civilization along the Nile River all used rivers. The rivers facilitated the transportation of goods, while the valleys held fertile agricultural land. Food surpluses meant that some people could be freed from the toil of food production and work at other tasks. Soldiers, tradespeople, craftspersons, merchants, priests, and road builders filled new roles, enabling the development of more complex societies. While each civilization developed its own distinct characteristics, the rivers continued to be instrumental in the social and economic successes of the groups.

Because of technology, people today are, for the most part, less dependent on rivers for transportation. Some societies, however, maintain very close ties to the water bodies on which they are located. Egypt is one such society. The arid, inhospitable climate of North Africa limits the uses of the land. As a result, the increasing population is dependent on the river for maintaining life. The contrasts are most striking: 99 percent of Egyptians live on just 4 percent of the land, that which borders the Nile River. These lands along the river have population densities approaching 1300 people/km^2, while the rest of the country is very sparsely populated. Egypt is the most **river-dependent region** in the world. Many of the problems this nation faces are related to the very limited amount of land that can be supported by the waters of the Nile.

1. Brainstorm a list of ideas, images, names, or places that you associate with Egypt. When you are finished, put a check mark besides those items on your list that relate to the history of the country; circle those that deal with things more modern. What does your list suggest to you about public awareness of Egypt?

2. **a)** Explain, in your own words, what is meant by a "river-dependent region".

   b) What are some ways in which rivers have affected development in another country of the world, such as Canada?

3. **a)** Why do you think the population of Egypt is concentrated in such a small part of the country?

   b) What problems might this create for the cities of Egypt? Check your answers to this question after you have completed this chapter. Rework your answers to reflect your new understandings.

4. Examine the photo in Figure 8.1.

   a) Describe the landforms and vegetation that are shown here.

   b) What activities are the people involved in? What tools are they using?

   c) Describe the way in which the land is irrigated in this part of Egypt.

   d) The term "labour intensive" is used to describe activities that require large amounts of human labour. "Capital intensive" refers to situations where money is spent to buy machinery to do the work. Which term more appropriately describes the work being done in this scene? Give reasons for your answer.

Figure 8.1 *Farming Irrigated Land*
Life in modern Egypt is a mix of the old and the new.

The Climate of Egypt

Egypt's climate is well known for its dryness. Any area that receives less than 250 mm of precipitation on average per year is called a **desert**. Using this criterion, all of Egypt can be classified as a desert. Yet, there is a snake-like ribbon of green, extending from the south to the north of the country. The Nile River has provided the water to maintain civilizations in this valley for thousands of years. To understand why this special river-dependent region exists, it is necessary to look at the factors that control the climate of northeastern Africa.

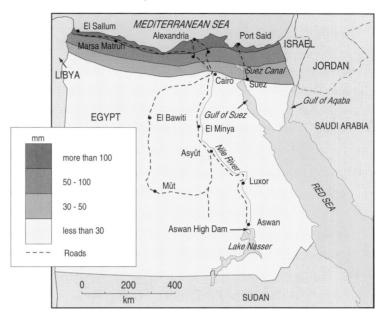

Figure 8.2 *Annual Precipitation in Egypt*
What is the most important factor creating the pattern shown on the map?

Figure 8.3 *Climate Statistics for Selected Cities in Northeast Africa*
The top row for each city gives the average monthly temperature in degrees Celsius.
The bottom row gives the average monthly precipitation in millimetres.

Location	J	F	M	A	M	J	J	A	S	O	N	D
Addis Ababa, Ethiopia (9°N, 39°E)												
Temp. (°C)	15.0	16.4	17.2	17.5	17.5	16.4	15.3	15.3	15.8	15.6	14.4	13.9
Precip. (mm)	13	38	66	86	86	137	279	300	190	20	15	5
Alexandria, Egypt (31°N, 30°E)												
Temp. (°C)	13.7	14.1	15.4	18.1	21.0	25.6	25.4	26.2	25.2	23.3	19.9	15.7
Precip. (mm)	49	24	10	3	2	0	0	0	1	6	33	56
Cairo, Egypt (30°N, 31°E)												
Temp. (°C)	12.3	13.5	16.4	20.2	24.2	26.7	27.7	27.6	25.3	22.7	18.7	14.0
Precip. (mm)	5	4	4	2	2	0	0	0	0	2	2	5
Entebbe, Uganda (0°N, 32°E)												
Temp. (°C)	22.8	22.9	22.3	22.9	22.8	21.1	20.5	20.7	21.1	21.6	22.8	21.6
Precip. (mm)	69	86	159	264	245	115	72.8	80	77	95	132	117
Khartoum, Sudan (16°N, 33°E)												
Temp. (°C)	23.7	24.7	28.0	31.5	33.7	34.1	31.7	30.4	32.0	32.3	27.9	24.9
Precip. (mm)	0	0	0	1	5	7	48	72	27	4	0	0

5. a) On a map of Africa, find each of the five cities for which climatic data is given. List them from farthest north to farthest south.

b) Compare the climates of the five places using a comparison chart. The points of comparison you should make are:

- latitude
- month and temperature of hottest month
- month and temperature of coolest month
- month and rainfall for wettest month
- month(s) and rainfall for driest month(s)
- annual rainfall

6. a) How does the annual average temperature of Addis Ababa differ from that of the other four stations? What are the reasons for this difference?

b) During which season does the rainfall maximum occur in Addis Ababa? in Entebbe? How do you explain these patterns?

c) Flooding of the Nile River in Egypt, before the damming of the river, occurred from June to September. Why did this flooding occur?

The Importance of the Nile River to Egypt

For thousands of years, the floodwaters of the Nile River deposited rich, fertile soil onto the fields of the Egyptian farmers. In fact, the cycle of living in the valley was controlled by the river. The floods came in mid-June and lasted to September. The fields were unworkable during this time, but the water was trapped and allowed to soak into the soil. Crops were planted and grew from October to February. Harvesting took place between March and June. The annual pattern of life of the people was linked to the rhythm of the river.

But this has always been a very dry area. Annual rainfall in Cairo averages 26 mm. From where did the floodwaters come? To answer this question, you must look to the source of the Nile River.

The Nile is the longest river in the world, about 6690 km in length. It has two sources deep in the Highlands of Central Africa. The White Nile begins as an outflow of Lake Victoria; the Blue Nile drains Lake Tana, located in the northern part of Ethiopia. The two rivers join near Khartoum, Sudan, and flow northward to the Mediterranean Sea, cutting a deep valley through the soft limestone of Egypt. A huge **delta** (a deposit of earth and sand that collects at the mouths of some rivers), some 240 km wide at its base, extends 160 km into the Mediterranean Sea. The annual heavy rains in the Central African Highlands give life to Egypt.

Figure 8.4 *The Sahara Desert*
This land contrasts sharply with the fertile lands of the Nile Valley. What conditions help to create the desert climate in Egypt?

As the floodwaters of the Nile rush out of the Highlands and towards the mouth of the river, they pick up many thousands of tonnes of **silt** (very fine earth, sand, and so on). In the valley of the Nile, in times past, the waters spilled over the banks of the river and onto the valley floor, where the farmers' fields were located. The much shallower water on the valley floor meant the water flowed less quickly and the silt settled out. When the floods receded, the silt was ploughed into the fields, enriching the soil. Thousands of years ago, the farmers built ditches and small canals to move water farther back from the river. For generation after generation, the cycle was unbroken. Historians have called Egypt "the gift of the Nile".

However, "the gift of the Nile" was felt within a very narrow band along the river. Beginning in the 1800s, the Egyptians began constructing irrigation systems to allow more distant fields to be watered year round. These systems used dams, canals, and reservoirs to capture and hold water so that it was available to farmers throughout the year. The amount of land that could produce food was expanded. These early irrigation schemes were hemmed in by the steep valley walls, limiting the amount of land that could be made to produce food to a maximum of a few kilometres on either side of the river.

Outside of the Nile Valley there are deserts. Two-thirds of Egypt's area is covered by the Libyan Desert, to the west of the river. To the east, between the river valley and the Red Sea, is the Arabian Desert. These areas are virtually uninhabited, with the exception of a few groups of Bedouins who have retained their traditional ways of life.

CHECKBACK AND APPLY

7. How is it possible for a major river like the Nile to exist in the dry desert climate of Egypt?

8. Trace the length of the Nile River on the map of Africa on pages A22-A23.

   a) About what percentage of its length is within Egypt?

   b) Suggest reasons why the Central African Highlands have greater amounts of rainfall than Egypt.

   c) Estimate the average width of the agricultural zone along the Nile. What confines the agricultural zone to a narrow band?

9. Figure 8.5 shows what a typical cross-section of the Nile Valley would look like. Use a series of similar sketches to explain how flooding of the Nile River fertilized the farm lands. Label your sketches.

10. a) In what ways might irrigation using canals, dams, and reservoirs change the way crops in the Nile Valley are grown?

    b) In what ways might such irrigation systems change the way of life for the people of the area?

Figure 8.5 *A Sketch Showing a Typical Cross-Section of the Nile Valley*

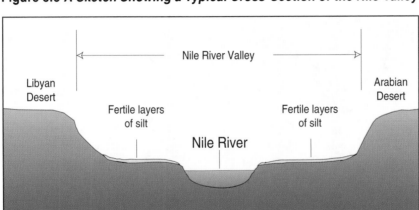

Modern-Day Egypt and the Nile

Today, as in the past, the Nile River sustains life in this region. In this century, a rapidly growing population necessitated the development of more land for growing food. The physical restrictions of the Valley meant that traditional solutions would not be able to keep pace with the population growth. New solutions were needed.

The Aswan Dam was the most spectacular of the solutions. This dam was built in the 1960s, in southern Egypt. It holds back the Nile waters and forms a huge reservoir called Lake Nasser. In addition to preventing the annual flooding of the river, it also yields a number of other benefits. The dam provides irrigation water to an additional 910 000 ha of land and has moved Egypt closer to self-sufficiency in food. There are now about 2.4 million hectares of farmland in the country. The reservoir is used for a fishing industry, and the dam generates enough electricity to meet a large part of Egypt's needs.

But the dam has also created new problems, most of which will be examined later in this chapter. Unfortunately for farmers, the lack of silt from flooding has meant that they must

Figure 8.6 *The Aswan High Dam*
What benefits did the Egyptians hope to gain by constructing this dam?

abandon traditional ways of farming. Chemical fertilizers are now used to increase the nutrients in the soil, in place of the silt. Crops are grown year round throughout the region using water brought through irrigation canals and ditches.

Although farming methods have changed, the way of life for Egyptians has remained largely the same. Rural people, called *fellahin*, still live in villages along the river as their ancestors did. Most farm small plots of land (average farm size is about 0.8 ha) or tend animals. Very often the land is rented from more prosperous landowners. The *fellahin* live in small, mud-brick houses, with thatched straw roofs. Both men and women work in the fields. Because farms are small, the land is intensively cultivated: this region is one of the most productive agricultural areas in the world.

Currently, 34 percent of the nation's labour force is engaged in agriculture, although agriculture contributes only 20 percent of the Gross Domestic Product. Egypt is the world's fifth largest cotton exporter and also produces rice, corn, wheat, beans, fruit, vegetables, and livestock (including cattle, water buffalo, sheep, and goats). Most of the fruit, vegetables, and animals are used within the country. In fact, in spite of the abundance of the Nile Valley, Egypt does not produce enough food for its own needs and must import from other countries, most notably the United States and European nations.

Not all Egyptians live in rural villages. Some 44 percent of the people now live in urban places. Cities, such as Cairo and Alexandria, have been growing rapidly as rural people have sought work in industries and businesses. (Manufacturing now employs about 15 percent of the labour force.) The cities are extremely overcrowded, and new arrivals often build makeshift squatter huts on rooftops or on public land. There is great pressure on the cities to expand outwards onto productive farmland.

Not everyone in the cities is poor. There are great contrasts in wealth, with quiet residential areas located near squatter settlements. Some Egyptians have lifestyles very similar to those found in European or North American cities.

11. **a)** Explain how the construction of the huge Aswan Dam prevented floods from occurring along the Nile River.

    b) The construction of the Aswan Dam created much hydro-electrical power for the country. Why was the dam built at Aswan instead of closer to Cairo where the greatest demand for power is?

12. Eroded silt is no longer being deposited in the fields during flooding. Where is the eroded silt upstream from the dam being deposited? What are the implications of this?

13. The Egyptians chose to build the Aswan High Dam in response to modern problems of maintaining food supply and the need for power. What other solutions might they have tried? Suggest three alternatives to dam building.

Tourism in Egypt

The warm, dry climate of Egypt and its wonderful relics of ancient times attract visitors from all parts of the world. The industry earned US$1.5 billion in 1987, up from just US$0.9 billion two years earlier. People come to view such widely famous features as the great pyramids and the Great Sphinx at El Giza. Ancient tombs in the Valley of the Kings and a variety of temples give travellers glimpses into the cultures and times of the past. The beautiful mosques, city walls, and gates attract people to Cairo.

But Egypt has more to offer than historic relics. It is a land where the ancient and the modern meet. Mud-brick villages stand near modern steel and glass buildings; TV antennae protrude from the roofs of straw thatched homes; the *fellahin* tend their small farms using archaic tools while some of the most modern military vehicles in the world rumble past. The mix of times and ideas is exciting.

A large part of the distinctiveness of the *fellahin* way of life comes from the religious beliefs of the people. About 90 percent of Egyptians are Muslims; the Islamic religion deeply influences the day-to-day activities of Muslims, including family and social relationships, business activities, and government of the country. Religious duties require praying five times a day, fasting, and giving alms to the poor.

Figure 8.7 *Feloukas on the Nile near Aswan*
This is a traditional design for travel on the Nile. What makes it a good form of transportation?

The spectacular beauty of the land also draws tourists. They come to see the green splash of the Nile Valley as it contrasts with the arid lands nearby. The Nile delta boasts white sand beaches — a sun worshipper's paradise. Mount Sinai and the **wadis** (ravines) nearby seem to change shape and colour as the sun travels across the sky. In total, Egypt has plenty to offer tourists looking for new experiences.

One of the most relaxing ways to see the Nile Valley is on a cruise boat. The trip from Luxor to Aswan takes about three days, and includes several stops so that passengers can explore the tombs, temples, and burial grounds of ancient royalty. Scenes on the walls of these ancient structures show that the lives of the *fellahin* have changed little over the centuries. Undoubtedly, the cruise boat will pass a *felouka*, a traditional sailboat of the Egyptians. Neither the *felouka* nor the cruise boat can get upstream of the Aswan High Dam.

CHECKBACK AND APPLY

14. What are the major attractions for foreign tourists to Egypt?

15. a) The ancient tombs and temples of Egypt generate a good deal of tourist income. What conditions have allowed such ancient structures to remain so well preserved?

b) In what ways might tourist activity lead to damage and destruction of ancient structures?

c) How might damage to historic sites by tourists be minimized?

d) What are other threats to the monuments of Egypt?

16. What aspects of tourism in Egypt are related to the fact that Egypt is a river-dependent region?

17. List five tourist activities you would suggest to the following people who are planning two-week vacations in Egypt.

a) a single, 25-year-old who has just completed university

b) a couple in their mid-40s travelling with their two school-aged children

c) a recently retired auto worker

Give reasons to explain your choices of activities.

No tour of Egypt would be complete without a visit to Cairo. The capital city bustles with cars, buses, and people. It is a mixture of new and ancient buildings and contains many parks. Street vendors are a common sight. Many people in the city depend on the tourist trade, often by trying to make a living selling drinks and snacks or other small items to tourists on the streets. Unemployment and a lack of adequate educational opportunities are big problems in the city. Nevertheless, Cairo remains the hub of northeastern Africa and is the centre of the country's travel routes, economic decision making, government, culture, military training, entertainment, history, and education.

Figure 8.8 *A Close-Up View of the Great Pyramid*
Estimate the size of each of the blocks that make up this pyramid. Use the people as reference. How do you think that the ancient Egyptians built these structures?

Famous Tourist Destinations

Cairo and the Pyramids

The ancient pyramids and the Great Sphinx are only 15 km west of Cairo. There are three pyramids at El Giza. The largest is about 150 m high and covers an area of four city blocks. In total, there are 80 Egyptian pyramids, which are believed to be the burial chambers of ancient rulers. One kilometre away from El Giza is the the Sphinx, with the body of a lion and the head of a man. These structures are visited by thousands of people every year. They reach the site by a paved road from Cairo. On this road, it is quite common to see people sweeping sand from the roads using brooms. The sand is constantly blown onto the road from the desert. If left to accumulate, it would make driving hazardous.

Figure 8.9 *Scenes in Cairo*
(a) These boys are making and selling fruit drinks.

(b) An Aerial View of Cairo

CHECKBACK AND APPLY

18. a) Locate Cairo on the map on pages A22-A23. Describe its location:

- within Egypt;

- in relation to the Upper Nile and the Nile delta;

- in relation to the Red and Mediterranean seas.

b) Before the Suez Canal was completed in 1869, goods travelling between the Red and Mediterranean seas were **transshipped** at Cairo. Transshipping involves the unloading of goods from one form of transportation and loading onto another. Explain why this would have led to the development of a city at Cairo. Give at least three reasons.

c) Cairo lost the transshipping business when the Suez Canal opened, yet it continued to prosper. Why?

19. What does the fact that people sweep the roadways tell you about employment in Egypt?

Egypt

THE REGION AT A GLANCE

Area	1 001 425 km²
Population (1990)	54 705 700
Population density	54.6 persons/km²
Urban population	44%
GDP* per capita (1989)	US$700
Population growth	2.5%
Birth rate	34/1000
Death rate	10/1000
Life expectancy	
—males	60 years
—females	61 years
Number of Egyptians working abroad	2 500 000
Exports	
—value of	US$2.5 billion
—major partners	U.S., European Community, Japan
Main industries	textiles, food processing, tourism
Imports	
—value of	US$10.1 billion
—major partners	U.S., European Community, Japan

*Gross Domestic Product

In addition to the ribbon of green along the Nile, there are isolated places in the desert where water comes close enough to the surface to sustain life. These places are called **oases**. (The plural of *oasis* is *oases*.) Water comes from beneath the surface to a low place where it can be used by people and animals. Archaeological evidence shows that many oases in the past had large lakes, and that the people who lived there grew, among other things, grapes and wheat.

Water in an **aquifer** (a porous layer of rock that channels the water along under the ground) flows slowly, at about 6 m per year. Excessive pumping for large-scale irrigation is not practical because of this slow rate, and oases communities are restricted from expanding. Nevertheless, these isolated oases settlements are home to a small percentage of Egyptians and provide important stopping places for travellers.

Tourists usually reach desert oases by truck. Many of the desert roads are no more than marked tracks, while some parts of the more travelled roads are paved. Camel caravans can still be seen in the desert, and the camel remains an important means of transportation because of its ability to withstand the rigors of the desert.

(a) An Oasis Village

Famous Tourist Destinations

A Desert Oasis

The appearance of date palms indicates the presence of an oasis. In some oases, there is surface water, but most depend on wells. A few oases are large. The oasis at El Fayum, where large quantities of fruit and vegetables are grown for the Cairo market, covers an area the size of a typical city of 2 to 3 million people.

Oases residents depend on agriculture for their livelihood. Irrigation permits the growth of crops such as millet, rice, vegetables, and date palms, the fruit of which are sold for export. Goats, sheep, and cattle may be raised using hay and other forage crops. Many of the old methods of farming — hoes, sickles, and cattle-driven sledges for threshing grain — are still in use, together with some modern machinery, such as water pumps and pesticide sprayers.

The flat-roofed houses of oases are made of sun-baked mud bricks and rarely have windows as a measure to keep the interiors cool. There is often an inner courtyard where a mud-brick oven is used for baking.

CHECKBACK AND APPLY

20. Compare ways of life of people who live in the Nile Valley with those who live in oases. Use a comparison chart and consider these points:

    - location
    - size of settlements
    - economic activities
    - transportation
    - potential for growth

21. Where do you suppose the water that is received at oases comes from, since it does not fall as rain in the surrounding areas?

22. a) Describe the potential for growth and development of oases in the desert.

    b) In what ways does this description of desert oases reinforce the idea that Egypt is a river-dependent region?

Figure 8.10 *Scenes at an Oasis*

(b) Irrigated Farmland and Date Palms

(c) A Well

Tourism Issue: Cultural Conflict

To many traditional Egyptians, the flood of tourists is not good for the country. Tourists bring with them values that are quite different from their own. It is difficult for some Egyptians not to be influenced by the obvious wealth and prosperity of the visitors.

Egyptians are already being influenced by American and European cultures. Television programs beam into their homes the latest events from around the world. Advertisements shout about the newest fast foods and chic clothes. Lifestyles previously unimaginable to the people bounce across the little screens in front of them. The images are of people living in absolute luxury.

Critics argue that Egypt has already become too North Americanized. Coca-Cola, VCRs, and blue jeans are not wanted so much for what they are, but for what they symbolize — the "better" life. Some people have welcomed this form of modernization, claiming that it motivates people by encouraging them to set goals and achieve financial success. But a large part of the country's *fellahin* see their way of life threatened by the rapid change. This is a culture that has endured for thousands of years virtually unaltered, and now it faces change of huge proportion.

Tourists visibly represent the culture of outsiders. It is they who make direct contact with the people of Egypt and it is their values that challenge traditional ways of life. Over time, Egyptians may begin to question basic beliefs and behaviours, such as what constitutes appropriate male and female roles in society. Since the beliefs of Egyptians are strongly linked to the Islamic religion, questioning of this type has to be viewed as undermining the religion of the people.

CHECKBACK AND APPLY

23. **a)** What characteristics of foreign visitors would you expect Egyptians to notice most readily? Why?

    b) In what ways might the interaction of two different cultures lead to changes in the cultures? Explain your answer.

    c) How might you decide what aspects from another culture are good?

    d) Identify two or three aspects of another culture that you have adopted. Explain why you adopted them.

    e) What does the fact that you adopted aspects of other cultures tell you about your culture?

24. **a)** The Egyptian culture has remained relatively the same for many hundreds of years. What does that indicate about the values of the culture?

    b) What aspects of foreign cultures might Egyptians find objectionable? Why?

    c) In what ways might the influence of foreign cultures be considered to be a problem?

    d) Suggest five ways in which the Egyptians might benefit from the influence of other cultures.

TOURIST GUIDE

- The official language of Egypt is Arabic. People who are educated and hold positions of responsibility usually also speak English or French.
- The work week is from Saturday to Thursday. Friday is the Muslim day of rest.
- Five times a day Muslim criers call out to the faithful to fill the mosques to pray. People remove their shoes before entering a mosque. Men and women sit in separate areas.
- Electric current in Egypt is 220 V, rather than the 110 V used throughout North America. Hand appliances may not work or will need adaptor plugs.
- Tap water in cities is usually chlorinated and safe to drink. In rural areas, it is advisable to drink bottled water.
- Many visitors who stay in Egypt for more than a week come down with "Pharaoh's Revenge" — diarrhea.
- Protect yourself from the heat and sun in Egypt. Headaches, dizziness, and nausea are signs that you have lost too much moisture and should get out of the sun.

The Future of Egypt

Egypt is a country very much focused on the Nile River. Ninety-six percent of its people and almost all of its economic activity are confined to this narrow valley. Many of the problems the country faces are a result of this dependence on the river.

Perhaps the most important problem for Egypt is the rapid population growth rate. The population is expected to reach 73 million by the year 2000, double the 1975 figure. For a country to have a large population does not necessarily mean that there is a problem, but when a country's resources are unable to meet the reasonable expectations of its people, then the country is overpopulated. The problem of overpopulation is seen most urgently in food supplies. Quite simply, since Egypt cannot expand its foodlands, it must import up to 60 percent of its food requirements. The money spent on imported food leaves the country and is, therefore, unavailable to stimulate other types of economic activity within the country.

Figure 8.11 *Population Pyramid for Egypt*
A high proportion of Egyptians are below the age of 15. What are the implications of this?

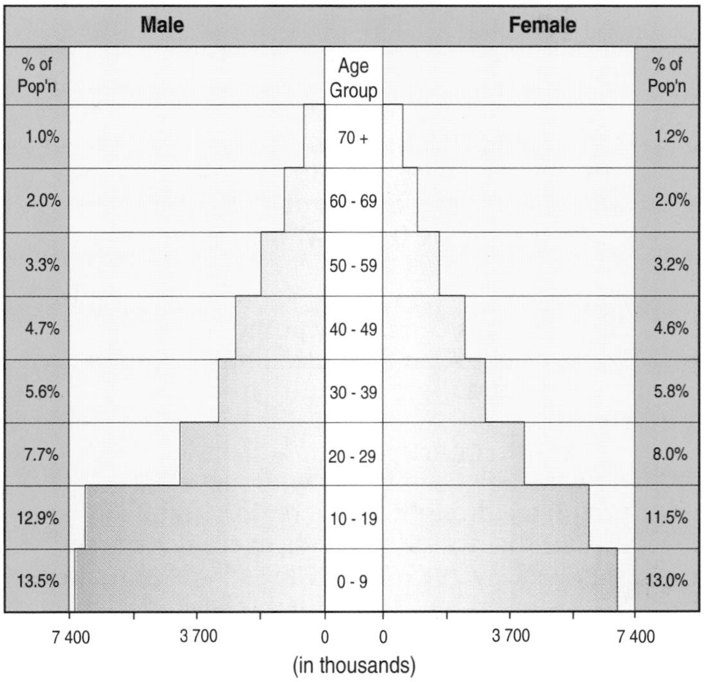

Male		Female	
% of Pop'n	Age Group		% of Pop'n
1.0%	70 +		1.2%
2.0%	60 - 69		2.0%
3.3%	50 - 59		3.2%
4.7%	40 - 49		4.6%
5.6%	30 - 39		5.8%
7.7%	20 - 29		8.0%
12.9%	10 - 19		11.5%
13.5%	0 - 9		13.0%

7 400   3 700   0   0   3 700   7 400
(in thousands)

Source: New World Almanac

To make up for the lack of domestic **capital** (investment money within the country) to stimulate the economy, Egypt has accepted foreign aid. In the past, this assistance has been used for such projects as modernizing Cairo's telephone and sewer systems and deepening the Suez Canal. Most foreign aid comes from the United States, but Egypt has also accepted assistance from other Western countries, other Muslim countries, and Communist countries. Unfortunately, much of the foreign aid is given in the form of loans which must be repaid. Because of this, Egypt has a huge foreign debt. In 1989, it owed other countries over US$45 billion. Repayment of this debt severely restricts the modernization programs that the country can undertake in the coming years.

It had been hoped that the Aswan Dam would give the country an economic boost. Many of the benefits have been realized — more irrigated land and a fishing industry in Lake Nasser, to name two — but some unexpected problems have limited the beneficial effects of the dam. One problem is the increase in **salinity** (saltiness) of the soil due to irrigation. Water evaporates quickly from fields in the intense heat of this climate. Unfortunately, the minerals in the water do not evaporate, but accumulate in the soil. Excessive amounts begin to poison the crops and reduce the productivity of the fields. When the soil was being flushed by annual flooding before the construction of the Aswan High Dam, salinity was not a problem. It now threatens to reduce food supplies even further.

Construction of the dam has also led to higher incidences of **bilharzia**. This disease is caused by worms that live in moving water. The worms enter humans through the skin while they are bathing in the water; sufferers grow gradually weaker and incur damage to their hearts, lungs, and livers. People suffering from chronic cases can barely put in more than three hours of work, so the nation has lost many millions of dollars annually from lost working time.

Another unexpected problem with the Aswan High Dam is low water levels. During a prolonged drought in the region in the mid and late 1980s, water demands exceeded water supplies. By 1987, Lake Nasser was at its lowest level since it had been filled. These low levels threatened Egypt's management of its only water resource and reduced the amount of hydro-electric power that could be generated.

In spite of these problems, there have been some success stories in Egypt. The country's manufacturing industries have shown substantial growth in recent years and continue to be strong. Over 2.5 million workers are employed outside of the country, primarily in the oil industries of other Arab countries, and the pay cheques sent home to Egypt improve the country's foreign currency situation. The country continues to play a critical role in the politics of the Middle East, being both a Muslim country and a country that maintains close ties to the West. Overall, the future is uncertain, but the potential for success exists.

CHECKBACK AND APPLY

25. Foreign aid carries both advantages and disadvantages.

a) Brainstorm lists of advantages and disadvantages of foreign aid for Egypt.

b) If you were a government official in Egypt, would you accept foreign aid from the United States? Give good reasons for your decision.

26. a) Use a flow diagram to illustrate how salinity increases in irrigated lands.

b) What might be some ways of reducing the amount of salinity in fields?

27. One solution to the problems created by the Aswan High Dam would be to dismantle it.

a) How could this be achieved safely?

b) What would be the short- and long-term consequences of such action?

c) Why do you think this action has not been seriously considered?

d) What other strategies might be used to overcome the problems caused by the Aswan High Dam?

28. Many "solutions" create different problems. Assess the pros and cons of the Aswan High Dam in terms of its environmental and economic effects.

Chapter Review

- Egypt can be defined as a river-dependent region because the life and economy of the country are influenced by the river.
- The Nile is fed by rains which fall in the central part of Africa; the water supports life along the length of the river.
- Oases are the only settlements outside of the Nile Valley.
- Agriculture is one of the most important activities along the Nile and helps maintain traditional ways of life.
- Most of the country's problems stem from the limited land and resources that are available for development.

Vocabulary Review

aquifer	river-dependent region
bilharzia	salinity
capital	silt
delta	transship
desert	wadi
oasis	

Thinking About . . .

EGYPT AS A RIVER-DEPENDENT REGION

1. **a)** List evidence that shows that Egypt is a river-dependent region.

   b) Describe some ways in which this influences tourism in the region.

2. Compare the way of life for farmers before and after the construction of the Aswan High Dam. Use a comparison chart and consider these points for comparison:
   - location
   - factors that influence agriculture
   - lifestyle
   - problems

3. In recent years, political instability in the Middle East has threatened the security of tourists in the area.

   a) What effect might such Middle East unrest have on tourism in the area?

   b) What actions might the Egyptian tourist industry take to reduce the negative effects of political unrest in the region?

4. In 1989, Canadian foreign assistance to Egypt amounted to C$29.74 million. Imagine that you are in a position to decide how the aid to Egypt is to be used over the next few years. Identify three programs in which you would choose to invest in order to create the greatest long-term benefits to the Egyptian people. In each case, carefully explain what you would do, and what the benefits of the program would be. Consider what people, equipment, and supplies would be needed, and from where these would come.

c) In general terms, what is the effect on an economy when imports are greater than exports? when exports are greater than imports?

d) Complete this table in your notebook:

Year	Imports — Exports	Percentage Imports of Exports

For each year, calculate the difference between imports and exports and enter the difference in the middle column. Find out what percentage exports are of the import values. Do this by applying the formula:

$$\frac{\text{Imports}}{\text{Exports}} \times 100$$

What do these two additional calculations tell you about Egypt's trade pattern?

e) Suggest methods that Egypt could use to try to improve the trade situation.

5. Figure 8.12 compares the value of imports and exports from Egypt every ten years, beginning in 1955.

Figure 8.12 *Value of Imports and Exports for Egypt*

Year	Imports (US$ millions)	Exports (US$ millions)
1955	65	51
1965	177	114
1975	603	210
1985	4880	1800

Source: Book of Vital World Statistics

a) Describe the changes that have taken place over the time period reported in the table.

b) Suggest reasons to explain the changes in the values of imports and exports.

Atlas Activities

1. a) Look at the maps and satellite image on page A15. Which of the three illustrations has the largest scale? Explain how you came to this conclusion.

b) Construct and complete a table that summarizes the advantages and disadvantages of each of the three illustrations.

2. a) Use evidence from the maps on page A15 to show the importance of the Nile delta to the people and economy of Egypt.

b) Explain what causes areas near the coast of the Nile delta to be less valuable for agriculture.

3. Refer to the maps of Bangladesh and the Tigris and Euphrates Basin on page A19. In what ways are each of these regions similar to, and different from, the Nile and its delta? In your answer, consider the nature of the land through which the rivers flow, the vegetation and climate patterns, and the distribution of the settlements in the areas.

Making Choices

Egypt's economic problems stem from the complex interaction of many factors, some of which are listed below:

- a lack of natural resources
- a rapidly increasing population
- the fragmentation of farmland
- the increase in salinity of irrigated soils
- the increase in waterborne diseases
- the expansion of cities, roads, and other developments onto scarce agricultural lands

Imagine that you are an official of the Egyptian government. The government is determined that Egypt will eventually become self-supporting. You have been asked to suggest strategies that would help to improve the economy and help Egypt to become self-sufficient. Suggest one strategy to tackle each problem listed above.

a) Set up a table with the following headings:

- Problem
- Cause of the Problem
- Strategies for Improvement

b) Fill in the table with a summary of your ideas.

LOOKING BACK AT EGYPT

 ## *Further Explorations*

1. As a visitor to Egypt, you would notice many special features. These would include:

   - the traditions of the Islamic religion
   - the historic structures, such as the pyramids
   - other aspects of culture, such as the special foods, crafts, social traditions, language, and music

   Conduct a detailed study of one aspect of the Egyptian heritage and culture that really interests you. You could present your findings in one of the following forms:

   - an essay
   - a wall or table display with explanations
   - a demonstration with explanations
   - a video program
   - another method approved by your teacher

2. Attempts have been made in a variety of dry areas in the world to make desert areas useful for agriculture. These include:

   - huge plans for water transfer from wetter areas
   - using underground water sources
   - planting trees to form windbreaks
   - spraying the sand surface with oil to prevent evaporation

   a) Investigate and report on two such programs that have been proposed or implemented in dry parts of the world. Include:

   - what the problem is or was
   - what was done or proposed
   - the results, or the reasons why the project was not implemented

   b) Evaluate the potential of such schemes for Egypt. Be specific in your comments.

3. Investigate the construction and design of the pyramids at El Giza. Your report should contain information about:

   - the reasons for their construction
   - the construction materials, including information about distances these materials had to be moved
   - how materials were transported and fitted into position
   - the dimensions, design, and alignment of the structures, and the significance which is attached to this

   Use a map, diagram, and/or a model to accompany your answer, if you wish.

9

Japan
A POLITICAL REGION

Figure 9.a *A Street Scene in Tokyo*

Figure 9.b *The Golden Pavilion, Kyoto*

Figure 9.c *A Rice Field Near Osaka, Japan*

Figure 9.d *Nikko National Park*

Figure 9.e *Vehicles Ready for Shipping*

STARTING OUT

1. Describe the location of Japan in relation to:
   - other countries of Asia;
   - the Pacific Ocean;
   - North America;
   - Europe.
2. List three economic activities suggested by these photos.
3. Comment on the contrasts that you observe in Japan based on these photos.
4. How is Japan's location an advantage and a disadvantage for tourism?

LEARNING DESTINATIONS

By the end of this chapter, you should be able to:
- summarize the strengths and weaknesses of Japan's location;
- describe aspects of the Japanese people's culture and way of life;
- recognize that the Japanese see themselves as people separate from the rest of the world;
- give reasons for the rapid expansion of Japanese trade and Japan's increasing wealth;
- identify tourist attractions in the country;
- recognize challenges faced by Japan as it moves toward the twenty-first century.

Introduction

Some time during 1986, a remarkable event took place, almost unnoticed by people around the world. During that year, Japan — a nation whose economy was devastated during World War II — became the world's largest lending nation. At almost the same time, the United States became the world's largest borrower of foreign money. This was a spectacular change in roles, since the United States was long accustomed to having the world's strongest economy. The seven largest banks in the world are now owned by Japanese companies. (Seventeen of the top twenty-five banks are Japanese owned.) What appears to many people as a great irony is that the United States owes a high proportion of its debt to Japan.

What circumstances led this small island country to take a leading role in the global economy? What conditions allowed it to climb out of the poverty and destruction of the war and to grow economically at such an outstanding rate? Many factors contributed to this change, but one of the most important was Japan's sense of itself as a nation. The Japanese people see themselves as separate from others. They view contributing to their society as a responsibility of every citizen. To the citizens of Japan, being Japanese is the most important fact of their lives. Propelled by this nationalistic and cultural spirit, the economy of the country rebounded after World War II.

The sense of uniqueness held by the people of Japan is reflected in the policies of the democratically elected government. Many government actions over the years have restricted the influence of others on Japanese society. In a number of ways, Japanese society is "closed", preferring to look inward, rather than outward to the world. This self-concentration, socially, culturally, and politically, has contributed greatly to the distinctive characteristics that define Japan as a **political region**.

CHECKBACK AND APPLY

1. Make a concept web to show what you already know about Japan. A concept web links words and facts about a topic. The web should begin like this:

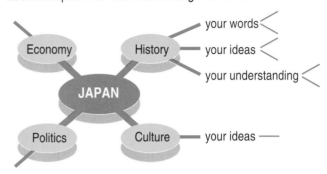

   Include at least 20 ideas in your concept web.

2. **a)** For what reasons can Japan be defined as a political region?

   b) What are some attractions political regions might have for tourists?

3. Given what you already know about Japan, write three questions that you would like answered in this chapter. Some words that you might use to begin your questions are:

   - Compared to Japan...?
   - What are three ways to explain...?
   - Why did...?
   - What if...?

   Check your questions as you work through the chapter to see which ones have been answered. How can you find answers to your unanswered questions?

Japan's Location in the World

The Japanese **archipelago** (a group of many islands) forms a long, gentle arc off the northeastern flank of Asia. There are about 4000 islands in total, the largest being Honshu. The northern islands are about the same latitude as Montreal and the southernmost point would compare with Miami's latitude. The Sea of Japan separates this country from the rest of Asia.

The islands of Japan are the tops of submerged mountains, part of chains or ridges that thrust up from the ocean floor. During the creation of these mountains, complex folding and faulting of the Earth's crust occurred, accompanied by volcanic activity. These islands are part of the "Ring of Fire" that runs most of the way around the Pacific Ocean. Seventy-seven of Japan's volcanos are considered to be active. Earthquake tremors are a fact of life in this region.

Japan's physical separation from Asia has been important in its development. For most of its history, Japan has had few serious threats of invasion. Fleets and armies would have had to cross the Sea of Japan, which has a reputation for sudden storms and dangerous conditions. This reputation did not, however, stop Japanese priests and merchants from venturing into the Asian countries. Often they returned with new ideas and new knowledge. The Sea acted as a filter, protecting the Japanese from the conflicts and turmoils of the Asian peoples, but allowing the benefits of the great Asian cultures to pass through.

The island location helped to isolate the Japanese from outside influences during the European exploration and exploitation of the Pacific. Between 1639 and 1854, the government shut the country off from the rest of the world. The government was so determined that Japan would remain uncontaminated by outside ideas that shipbuilding was restricted to vessels that could not go far from shore, and individuals who tried to leave the country were executed. When restrictions were finally lifted in the nineteenth century, the economic and social structures of the nation were primitive compared to many other countries. Although economic and social conditions have radically changed since then, the sense of being separate from others remains strong.

Figure 9.1 *Rocky Coastal Scene in Japan*
What advantages does the fact that Japan is an island have for the Japanese people?

CHECKBACK AND APPLY

4. Describe the advantages and disadvantages of Japan's location for each of the following groups of people:

   - importers of raw materials
   - exporters of consumer goods
   - Japanese tourists
   - foreign tourists to Japan

5. **a)** List some advantages and disadvantages of being culturally isolated from outside influences.

   b) Identify some influences, traditions, and customs from other cultures that have been accepted into Canadian culture

   c) Is it possible for a culture to be too much influenced by people of other cultures? Give your opinion on this question and support it with examples and evidence.

Japan's People and Culture

To outsiders, the Japanese appear to be a very **homogeneous** people (much like one another). There is remarkably little ethnic diversity (differences), with only 900 000 foreigners living in a country of over 125 million. Most of these foreigners are Koreans, who are deliberately segregated from Japanese society. Between 1980 and 1985, only 37 000 people were naturalized (made citizens) of Japan: in the United States, more than that number of people are made citizens each month. The Japanese word for non-Japanese people is *gaijin*, meaning "a person from outside". *Gaijin* living in Japan must accept that they will never be part of Japanese society. In Japan, unlike in many other places, there is only one entry to Japanese society, and that is by birth. If you are not born Japanese, it is extremely difficult to become a Japanese citizen.

The Japanese population is not, however, truly homogeneous: there are divisions along ethnic lines. The Japanese as a distinct ethnic, cultural, national group are descended from waves of immigrants who, over thousands of years, came to Japan from Asia. Each group made its own contribution to the Japanese culture. The kimono, for example, came from the Chinese. But ethnic differences are subtle and have been down-played by the business leaders and politicians. The island nation displays a strong sense of cohesion (togetherness) and shared purpose. It is one of the country's greatest strengths.

Conformity and **consensus** (general agreement) are important characteristics of Japanese society. Because the country is mountainous, the population inhabits only a small portion of the land; overcrowding is a way of living in Japan. Fitting in with others and going along with the group's decisions are skills that the Japanese learn early and practise throughout life.

Figure 9.2 *A Scene in Rush-Hour Tokyo*
Traffic congestion is a way of life for the Japanese. Seventy-five percent of the land is mountainous and largely uninhabitable.

Figure 9.3 *A Typical Japanese Family*
What evidence is there in this photo of respect for traditional values?

Language, as well as geography, has contributed to the Japanese sense of separateness. No other major language is so restricted to a particular geographic area as is the Japanese language. Largely due to the isolationist policies of the government up to the middle of the nineteenth century, the Japanese language was unknown outside of the country. Although immigration since then has spread the language, there are relatively few non-Japanese people who are fluent in it.

Japanese society is, however, very receptive to new ideas that it views as leading to improvements. The music, sports, technology, and foods from other places have enriched Japanese society. You can, for example, hear American country-and-western music in a bar in Tokyo, eat at an Indian restaurant and be served by Japanese waiters in turbans, or watch British sit-coms with Japanese sub-titles. This dynamic mix of foreign and Japanese ideas and influences has given the society an interesting look, although, in many ways, the real nature of the society has not changed.

CHECKBACK AND APPLY

6. a) In what ways has the Japanese society benefited from being open to new ideas?

b) In what ways has the society benefited from being isolated?

c) Given a choice, would you prefer to live in an open or closed society? Explain your answer.

7. Suggest reasons why only a small number of people are given citizenship in Japan at any time.

8. Some countries have a saying: "The squeaky wheel gets the oil." In Japan, the saying is: "The nail that sticks out gets hammered down."

a) Explain what each of these sayings means.

b) Explain how the second saying reflects the social and cultural characteristics of Japan.

Japan
THE REGION AT A GLANCE

Area	377 835 km²
Population (1990)	123 642 000
Population density	327.2 persons/km²
Urban population	76.7%
GDP* per capita	US$15 600
Population growth rate	0.4%
Life expectancy	
—males	76 years
—females	82 years
Inflation rate (1989)	2.1%
Unemployment rate	2.3%
Defence spending	1% of GDP*
Share of GDP*	
—primary industries	3.7%
—secondary industries	37.8%
—tertiary industries	58.5%
Households with colour TV	99.1%
Main industries	electrical and electronic equipment, automobiles, machinery, chemicals
Main exports	machinery, motor vehicles, consumer electronics
Main imports	manufactured goods, fossil fuels, foods, raw materials

*Gross Domestic Product

The Economy of Japan

Perhaps nothing shows the Japanese sense of commitment better than their approach to business. All levels in society agree that achieving the maximum economic advantage is the most important goal. The government facilitates this through low taxes, low interest rates, high levels of savings, and high capital accumulation. There is little direct government interference in the economy; rather, there is a strong reliance on the private sector to make the best decisions. Capital (investment money) is readily available as the Japanese are some of the best savers in the world, routinely putting an average of 16 percent of their earnings into savings accounts. Taken together, these savings can be used to spur business development in Japan or to invest in ventures outside of the country.

The huge amount of available capital means that companies can take a long-term view in planning. Business activities do not have to realize immediate profits; the long-term success of ventures is of greater importance. Achieving the best overall level of return is seen as preferable to a quick, but small, profit.

The long-term approach to business is partly a reflection of workers' attitudes. People are usually employed for life and view the company's goals as their own. As a result, Japanese workers routinely put in work weeks that are 10-15 percent longer than those of Europeans and North Americans, as Figure 9.5 shows. Co-operation between workers and management is expected. Because workers know their jobs are secure, they are not concerned about the introduction of new technologies or labour-saving devices.

Figure 9.5 *Average Work Week for Selected Countries*

Country	Regular Working Hours	Extra Working Hours
Germany	32.1	1.7
Japan	39.6	4.3
United Kingdom	36.3	3.3
United States	35.6	3.6
Canada	36.7	0.9

Source: Book of Vital World Statistics

Figure 9.4 *Tomato-Picking Robot*
Japanese industries lead the world in using new technologies, such as robotics. Through image processing, this robot can differentiate between colours of tomatoes and leaves, and is programmed to pick only ripe tomatoes.

The quality that seems to give Japanese workers the greatest economic advantage is their ability to reach consensus. When a decision in the work force is made, everyone agrees to it and fully supports it. Decisions are made by general agreement, not by the majority, nor by those who hold the greatest power. Reaching consensus is not easy: it requires careful consultation with all involved. As a result, decision making is a slow process, but once reached, a decision is supported by all.

Japanese industrial leaders recognize that to maintain their competitive edge they must remain in the forefront of new technologies. One technique that has been successful for Japanese industries is the use of standardized **components**. Standardized components are parts with particular functions. Different components can be combined in different ways to manufacture different products. That is, the same components put together in different ways create new and different products. These standard components become custom-built products, at about half the cost of products made using regular techniques. This type of production is accomplished by computerized scheduling that enables different standard components to arrive at the right place and time so different products can be created on one production line. Using techniques of this type, Japanese businesses have reduced typical **turnaround times** from three years to six months. Turnaround time is the period between the design of a new product and its production for the market. Reduced turnaround times give firms a competitive edge because of time and cost reductions.

Japanese companies invest twice as much money in research and development of new products as does the United States. Japan is now the world leader in innovative technologies and applies for more patents than any other country in the world.

CHECKBACK AND APPLY

9. **a)** Describe the role of the Japanese government in business.

   b) Identify two different roles that government can play in business. In which countries does the government play these roles?

   c) Given what you now know about Japan's society, why do you think that the economic role played by the Japanese government is the best one for Japan?

10. How do people's savings make improvements in industries possible?

11. **a)** Give examples from your own experience of how decisions are made by majority, by consensus, and by authority.

   b) What characteristics does a group require before it can become good at reaching consensus?

   c) What are the advantages of consensus over such forms of decision making as rule by majority or rule by authority?

   d) What are the disadvantages of decision making by consensus?

12. Examine the following table which gives information about the number of patents applied for and registered in selected countries.

Figure 9.6 *Number of Patents by Country*

Country	Applications	Registered
Japan	440 219	105 905
United Kingdom	37 093	29 590
United States	109 625	57 889
West Germany	82 851	30 501
Canada	23 992	21 061

Source: Science and Technology Agency Statistics Canada

a) What is a patent?

b) What does the number of patent applications indicate about a country?

c) What advantages does a country gain from a high number of patent applications?

d) How might more research and development be stimulated in a country?

Case Study

Inside a Japanese Factory

Tomiko is a recent high school graduate employed on the assembly line at S.H.T. Electronics. In the morning, Tomiko is picked up at the commuter train station, along with thousands of other co-workers, by company buses. The bus takes her to the factory in a quiet suburb of Tokyo. Upon arrival, she clocks in and goes to her work station. As always, she wears a blue smock over her street clothes, the same as all other employees. The plant is spotlessly clean and bright.

Tomiko has been assigned to a team, which is called a section. Most of the people on the line are women, and most of the supervisors are men, although this is slowly changing. At 8:27 a.m. the day starts with light physical exercise, and at 8:30 a.m., each section holds a short meeting to discuss its goals for the day. If, as the day progresses, it appears that targets will not be reached, the supervisor of the section speaks individually to each worker. Those sections exceeding their goals receive cash bonuses.

There are three breaks in the day — a forty-minute lunch period and a ten-minute break in the morning and afternoon. At the end of the eight-hour day, the workers clock out and ride the buses back to the train station.

To keep spirits and loyalty high, the company organizes sports days and other company events. It offers lifetime employment and personal benefits, including medical care, a pension plan, and bonuses.

Much of the plant where Tomiko works is automated, with industrial robots performing repetitive tasks. The company was careful to ensure that no one became unemployed due to the introduction of these machines. Also, no jobs were lost when international competition from Korea led to the closure of one line.

CHECKBACK AND APPLY

13. a) In what ways are workers expected to conform to the company's policies and practices?

b) In what ways do workers benefit from being part of the company?

Figure 9.7 *Television Assembly Plant*
Industrial robots are used in many industries in Japan.

Japan's International Role

Japan's economic successes have been built on its foreign trade. In 1989, the country had exports of US$270 billion, 97 percent of which were manufactured goods. Its largest trading partner was the United States (34 percent of exports); the rest of its trade was split among Southeast Asian, Western European, Communist, and other countries.

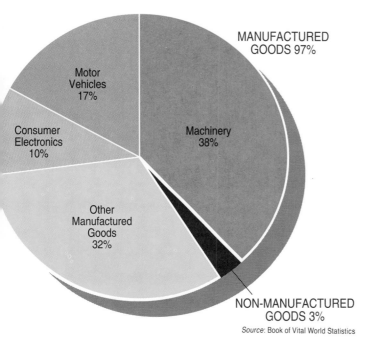

Figure 9.8 *Japan's Exports by Type*
Suggest reasons to explain why manufactured goods make up so much of Japan's exports.

The country's imports for 1989 totalled US$210 billion, giving a **trade surplus** (the amount by which exports exceeded imports) for that year of US$60 billion. Figure 9.9 shows imports by type. The United States was the largest source of imports, with 23 percent of the total. Other major trading partners were Southeast Asia, the Middle East, and Western Europe. Overall, Japan's economy grew by 4.8 percent in 1989, although the industrial sector posted a growth rate of 9.0 percent.

Figure 9.9 *Japan's Imports by Type (Percent)*

Imports	1980	1989	1993 (forecast)
food and raw materials	28	28	26
fossil fuels	50	30	24
manufactured products	22	42	50

Source: Book of Vital World Statistics

The large trade surplus means that the country is earning money from its exports. This money can be reinvested in the industries themselves, or in **foreign investments**, such as real estate or corporations in other countries. Japanese investment overseas is growing so rapidly that, by the turn of the century, the country's investors may control an estimated one trillion dollars (US$1 000 000 000 000) in assets in other countries. American interests are particularly targeted by Japanese investors because of the large, affluent market in the United States. The figures are staggering: over a million Americans will work for Japanese companies by the mid-1990s; 75 percent of Waikiki's beachfront hotels are owned by Japanese investors; Japanese investment firms are among the largest purchasers of U.S. Treasury bonds. One Japanese company, Nippon Telegraph and Telephone, is worth more today than IBM, AT&T, General Motors, General Electric, and Exxon combined!

Japan's interests in other countries has not been limited to the United States. Similar purchases of assets have been recorded in most other stable economies around the world.

Huge Japanese trade surpluses worry Japan's trading partners. The United States' trade deficit with Japan is a drain on the U.S. economy. This has prompted American demands that the Japanese should import more from the U.S. or establish more manufacturing facilities, especially for automobiles, in the U.S. In response to these demands, Toyota, Honda, Mazda, Nissan, Suzuki, and Subaru-Isuzu have built plants in the U.S. and Canada. Honda has, for example, established a plant in Alliston, Ontario, Toyota has established one in Cambridge, Ontario, and other companies have located manufacturing facilities in Europe. In the electrical machinery industry, another strong Japanese sector, 66 companies have established overseas subsidiaries. These Japanese-owned businesses in other countries contribute to employment in those countries and facilitate the transfer of high-tech methods and equipment.

Figure 9.10 *Japanese Business People Abroad*
What advantages does foreign investment give Japan?

But Japan's focus on itself, and the people's view that the society is unique, has led to criticism of the country. Critics argue that Japan has not played enough of a leading role on the international scene, given its economic importance. They argue that with wealth goes responsibility to others. One way of showing this responsibility would be to import more from other countries. Manufacturers in developing countries would then be able to compete for the large and prosperous Japanese consumer market. Japan has made some concessions on this front, but not significant ones.

Another way in which Japan could assume greater global responsibility is by taking a more active role in maintaining a peaceful world order. Japan could, for example, use diplomatic means to help solve conflicts between countries. It could also use its economic strength to encourage countries to act in ways that would create greater harmony among nations. Critics are optimistic that Japan is moving in these directions, but point out that the Japanese will have to change their historic sense of separateness in order to play these roles well. Japanese society will have to become more involved in international events: this "internationalization" may not come easily to the citizens of Japan.

CHECKBACK AND APPLY

14. **a)** Why was Japan able to create a trade surplus with many other nations?

    b) Why are other countries concerned about Japan's huge trade surplus?

    c) Japan has reduced its trade surplus somewhat in recent years, partly by setting up manufacturing industries in other countries. List at least two advantages and disadvantages of this from the host country's point of view.

15. Suppose you were in charge of industrial growth in a developing country in Africa, Asia, or South America.

    a) What would be your attitude towards Japanese foreign investment in your economy?

    b) What strategies might you use to ensure that your country received maximum benefits from the foreign investment?

16. Why is maintaining world peace crucial to Japan?

17. Do wealthy nations have a responsibility to help poorer countries? Write a one-page opinion statement on this topic, giving good reasons for your point of view.

Tourism in Japan

The apparent lack of interest in the outside world by the Japanese people is reflected in their travelling patterns. Only one in twenty-five Japanese travel outside the country, versus one in three for the British. When they do travel, the Japanese tend to prefer group tours, which limit their contacts with foreigners. The high costs of accommodations and travel within Japan discourage outsiders from visiting. The number of visitors (2.3 million in 1988) is equivalent to only 2 percent of the native population, compared with 25 percent for the United Kingdom and 60 percent for France. For the vast majority of Japanese, tourism does not increase their direct contact with foreigners.

Many Japanese people are reluctant to travel, and a significant proportion of workers do not even take all the holidays to which they are entitled. One reason for this is the extremely high cost of living and of taking a vacation. A holiday in Japan costs about three times as much as an equivalent holiday in North America because of high prices for train tickets, highway tolls, gasoline, and hotels. Figure 9.11 shows that long trips by Japanese families are not very common, unless visiting one's hometown, where accommodation may be arranged with family members.

Figure 9.11 *How Japanese Families Spend Their Summer Vacations*

Short Overnight Trips in Japan	54%
Day Excursions and Leisure Activities	47%
Return Visits to Hometowns	39%
Overseas Trips	3%

Source: World Tourism Organization

Note: The total is greater than 100% because many families take several different kinds of vacations.

The results of the prohibitively high cost of taking a vacation in Japan are also clearly shown in Figure 9.12.

Figure 9.12 *Travellers To and From Japan, 1987*

Country	To Japan	From Japan
Australia	42 891	215 562
Canada	58 286	311 448
China	69 561	577 702
France	37 185	611 796
India	27 442	46 355
Italy	20 880	384 837
South Korea	216 355	893 596
Switzerland	15 841	369 130
U.K.	148 302	297 200
USA	550 261	2 128 481
West Germany	53 543	588 615

Source: World Tourism Organization

In spite of the high cost of travelling, more and more Japanese are becoming interested in tourist travel. This is largely because of rising incomes and standards of living in the country. Between 1988 and 1989, travel agents reported that domestic sales rose by 6.6 percent and overseas travel by 13.3 percent. The stronger value of the yen compared to other currencies in recent years has made foreign travel somewhat more attractive than domestic trips.

For both foreigners and Japanese who do travel in Japan, there is a full range of accommodation available. In the major cities, there are modern hotels, many of which offer the kind of room and service available in the best of North American hotels. Less expensive hotels with single rooms and fewer luxuries are available in the commercial and industrial parts of major cities. In vacation areas, in addition to large hotels, you will also find lodges, where only a few guests are accommodated at one time. Here you can experience the traditions of the country more directly, such as sleeping on a futon (mattress) on tatami mats on the floor and being served food at a traditional low table. For economical travellers, there are travel hostels, where there is a minimum of service, but where you can meet other travellers in an informal setting.

A journey to Japan would not be complete without experiencing the Japanese railway system. Thirty-five percent of all travellers use the railway, in contrast to less than ten percent in North America. Fast and comfortable railway transportation is based on a system of some of the fastest commuter trains in the world. The "Bullet" or Shinkansen trains are powered by electricity and can travel at speeds of 240 km/h. They take only three hours to travel between Tokyo and Osaka, a distance of 520 km.

CHECKBACK AND APPLY

18. Why has tourism not been particularly effective in making the Japanese people more international in their perspective?

19. a) Describe Japan's balance of trade in tourism.

b) What factors have created the tourism balance of trade?

c) In what ways might the balance of trade in tourism change in the future? Give reasons for your answer.

20. a) On a blank map of the world, colour and label Japan. In another colour, label each of the countries listed in Figure 9.12.

b) Plot arrows on the map to show sources and destinations of tourists. Use one colour of arrows to illustrate the number of journeys to Japan, and a different colour to illustrate the number of journeys from Japan. Write the actual number of journeys along each arrow.

c) Suggest three reasons why more tourists leave Japan than visit Japan.

d) What strategies could the Japanese government employ to attract more foreign tourists?

e) If the Japanese government wanted to encourage Japanese people to spend more of their vacation money in their own country, how should they do so?

21. Using examples from the text, and from other sources if you wish, explain how:

a) a flourishing tourist industry benefits from a healthy domestic economy;

b) a healthy domestic economy benefits from a flourishing tourist industry.

One of the major obstacles to Japanese development has been the difficulty of convenient and rapid travel throughout the nation. To overcome this, the Japanese have built two undersea tunnels and a series of bridges linking the islands.

The first undersea tunnel in the world was built in 1944 to link the islands of Honshu and Kyushu. The Seikan Tunnel was completed in 1988, and links Honshu with Hokkaido to the north.

In April, 1988, the Seto-Ohashi bridges, linking Honshu with Shikoku, completed the long awaited connections between the four main islands.

Starting in 1972, the Seikan Tunnel involved 13.8 million person days of labour, 6.33 m^3 of concrete, 168 000 t of steel, and 847 000 m^3 of grouting (material forced into the cracks in the rocks to strengthen and waterproof them). The tunnel has a total length of 54 km, 23 km beneath the sea, and exceeds the length of the tunnel that links Britain and France. Some of the actual and anticipated benefits resulting from the construction of the Seikan Tunnel include:

- a reduction in ferry delays and sinkings due to bad weather;
- the attraction of industry and people to Hokkaido, and easier access to Hokkaido products in other parts of Japan;
- easier access to the Hokkaido market by Honshu industries;
- speedier transportation of fresh agricultural and fish products to Honshu, and increased volume of sales;
- increased passenger traffic and tourism.

Trains using this tunnel will transport automobiles and will have luxurious sleeper cars when travelling between Tokyo and Sapporo. It is anticipated that bullet trains will use this route at some time in the near future.

Famous Tourist Destinations

The Seikan Tunnel

Figure 9.13 *The Seikan Tunnel*

22. Constructing and using the Seikan Tunnel presented challenges which are not usually found in other tunnelling situations.

a) What problems can you see that would have to be overcome in building an undersea tunnel?

b) What additional safety features would you add because of the Seikan Tunnel's location on an earthquake zone?

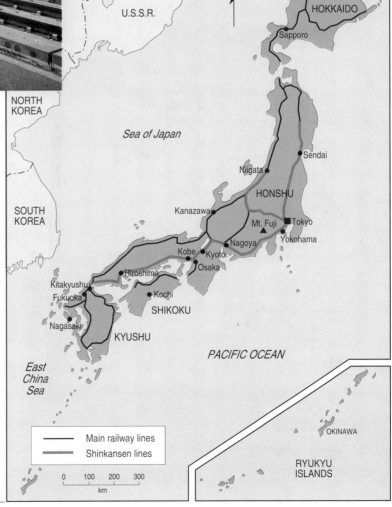

Figure 9.14 *Japan's Railway System*

TOURIST GUIDE

- The usual form of greeting in Japan is a bow, not a handshake. A bow signifies, "I respect your experience and wisdom." With someone of greater age or status, you bow low and long; with someone of less age or status than yourself, a brief bow will do.
- Always address a Japanese person by his or her last name, unless addressing a very close friend. Putting "San" after the last name is the equivalent of adding "Mr." or "Ms."
- Before entering a house, remove your hat, gloves, and shoes.
- Gift giving occurs very frequently in Japan. Gifts should always be wrapped in pastel colours; black and white are associated with funerals and red is not considered to be appropriate. It is not customary to open a gift in front of the giver, unless asked to do so. This is so the giver is not embarrassed by your reaction, or lack of it.
- The best times to travel in Japan are in the spring and fall. The summers tend to be humid and the winters grey and chilly.
- Tips are not expected in Japan. Take along small gift items to reward people who have given you good service.
- Visitors are advised not to drive in Japan, as roads are narrow and congested and signs are seldom in English. The costs of fuel and road tolls are high.

Tourists travelling to large cities almost anywhere in the world can expect to pay relatively high costs for food, accommodations, and transportation. In Japan, particularly in Tokyo, costs will seem staggering to the average tourist.

Almost one-quarter of the population of Japan lives in the Tokyo area, which is crowded and expensive. For example, land in downtown Tokyo sells for up to $462 000 per square metre, whereas in a place like downtown Toronto it sells for about $10 750 per square metre. As a result, hotels and restaurants are expensive, as land costs must be passed on to consumers. Most costs that tourists will incur are two to three times higher in Tokyo than in cities such as Washington, London, or Paris.

While tourists may feel the pinch of high prices, the Japanese must live with the sky-high costs. Most people who own homes cannot afford to live close to their places of work and have to commute long distances. Still, suburban houses are very expensive, and some people have taken out 100-year mortgages to make home ownership possible, passing the debt from one generation to the next until the house is paid for. A typical home in a Tokyo suburb, in which a family of four might live, would be made of wood, with a tiled roof, two stories, and a small garden. Inside, the house would have 86 m^2 of living space, including three bedrooms, a dining/kitchen/living room, and a storage room. Many families live in rental accommodations because they cannot afford to buy houses.

Nine cities in Japan have subway systems; in Tokyo the subway is about 200 km long. With 2.6 million commuters each working day, it is much more crowded than any North American system and is a good place for tourists to avoid during rush hours. By the time a commuter train reaches central Tokyo, the people are packed in tight. However,

Tourism Issue

The High Costs of Travel

subways provide a vital alternative to the polluted and traffic-choked streets of the city.

To reduce the expense of accommodations, some tourists spend a night in a capsule hotel. These have sleeping compartments, similar to those on a train, stacked two or three high. You can sit but not stand in a capsule. Each one is equipped with a TV, radio, alarm clock, and light. A shower room is at the end of the section.

Food is also very expensive by North American standards. For example, two medium-sized tomatoes might cost $6.25; one apple, over $1.00; and a cup of coffee, $3.50 to $5.00!

CHECKBACK AND APPLY

23. Imagine that circumstances, such as the transfer of a parent, meant that you and your family had to live in a Tokyo suburb for the next three years.

   a) What would you consider to be your greatest challenge and how would you overcome it?

   b) In what ways would your everyday life, and that of the other family members, change?

   c) List those things that you would probably like, and those that you would probably dislike, about your stay in Japan.

   d) How would your experience be different from that of a Canadian tourist who stayed for two weeks?

   e) In what ways do you think the experiences and feelings of a Japanese student coming to Japan under the same circumstances would differ from your own?

Figure 9.15 *A Section of Downtown Tokyo*
The downtown core of Japanese cities does not rise up as high as in North American cities. Can you think of the reason why skyscrapers are not allowed? These buildings consist of retail, commercial, and industrial operations.

Figure 9.16 *Homes in a Suburb of Tokyo*
These houses are valued at $600 000 to $800 000 each. The roof in the foreground is that of a factory. The Japanese do not divide cities into land-use zones, and industry, residential areas, and even farms are often seen in close proximity.

The Future of Japan

As it approaches the twenty-first century, Japan faces a number of challenges. Perhaps the most important is its growing international role. The country's rapid economic growth and huge wealth have disturbed established patterns, especially in the United States where people feel that the American way is being threatened. Greater global pressure will be put on Japan to assume a greater leadership role in industry, technology, and politics. If it fails to do so, its high level of foreign investment may be viewed with suspicion. But, taking on such international roles runs counter to the attitudes and values of traditional Japanese society. It will take major shifts in thinking patterns to make the Japanese nation more global in its perspective.

Social change will also challenge the Japanese people. Young people are questioning the practices of conformity and consensus. With increasing wealth and more leisure time, these people are developing interests that cause them to fall out of step with the established social order. Great changes are taking place in attitudes towards money, privacy, and duty.

The changing values of young people is just one aspect of cultural change with which the country will have to deal. As it becomes more economically linked to the Western world, Japan will have to respond to mounting pressures for change. Some of these pressures will come from within, from citizens who have travelled to or studied in other countries, from viewers of Western television shows that are carried on Japanese signals, from those critical of traditional ways of life.

Whatever criticism there may be of Japan, clearly it is a nation that is adaptable and capable. Its common purpose will see it through the stresses and pressures of being a world leader. Indeed, it has been said by many that the twenty-first century may well be the "Japanese Century".

CHECKBACK AND APPLY

24. **a)** List five to ten roles that wealthy countries play in the international scene. Your list might include economic, military, technological, or environmental aspects.

    b) Why would there be pressure on Japan to take on some of these roles?

25. Suppose Japan does not take an active international role in non-economic issues. Why might some nations view its foreign investment with concern? Explain your reasoning.

26. Will tourism in Japan improve over the coming years? In what ways might the industry change? Give reasons or evidence to support your opinions.

LOOKING BACK AT JAPAN

 # Chapter Review

- Japan's location allowed it to develop quite separately from other groups of people.

- Throughout the years, the Japanese learned how to use others' ideas and techniques to benefit their own society.

- The Japanese people have a strong sense of their uniqueness as a society. Conformity and consensus are valued in Japan. The group takes precedence over the individual.

- Japanese industries have benefited from the loyalty shown by employees; employees have benefited from a high standard of living and little unemployment.

- The success of the Japanese economy has allowed the country to become a world superpower, at least in economic terms.

- Japan's economic success has led to calls for its more active involvement in global affairs. This runs against many of the values held by the people.

- Tourism in Japan is hampered by high costs and a lack of interest in travel by the Japanese people.

 # Vocabulary Review

archipelago
component
consensus
foreign investment

homogeneous
political region
trade surplus
turnaround time

 # Thinking About...

JAPAN AS A POLITICAL REGION

1. **a)** List five factors that you think have contributed to Japan's rise to economic power. Rank them in order of their importance. Defend your ranking.

   b) What actions have the Japanese taken recently to ensure that their economy flourishes in the future?

2. In what ways has Japan's location benefited the country? In what ways has it harmed the country?

3. What can North Americans learn about business from the Japanese? Prepare a list of five actions that businesses in North America could undertake to make themselves more competitive with the Japanese. For each action, explain what would be involved and how it would benefit the business.

4. Examine the statistics on Japan's labour force given in Figure 9.17.

   a) What trends do the data show?

   b) What explanations can you offer to account for these trends?

   c) In what ways might these trends indicate changing attitudes and values of the Japanese people?

Figure 9.17 *Japan's Labour Force (Percent)*

Category	1985 (actual)	1993 (est.)	2000 (est.)
Primary	8.8	7.2	5.5
Secondary	34.3	32.1	31.0
Tertiary	56.9	60.7	63.5
Total	100.0	100.0	100.0
Total Number of Workers	58 070 000	61 340 000	63 110 000

Source: Japanese Board of Trade

5. a) What conditions restrict the growth of tourism in Japan? Which condition is the most important? Give reasons to justify your ideas.

b) What might the Japanese tourist industry do to try to stimulate more travel by the Japanese? Give three specific actions.

c) What might the North American tourist industries do to try to encourage more Japanese tourists to travel on this continent? Give three specific actions.

6. List five problems that you think will be important to the Japanese in the coming years. You might think about such topics as the economy, the environment, social conditions, energy, and the like. Explain why you have selected the problems for your list.

Atlas Activities

Refer to the maps of Japan on pages A16-A17.

1. a) Describe the distribution of population in Japan and identify the factors that have influenced this distribution.

b) In what ways is there a connection between the areas of dense population and specific kinds of agricultural production? Suggest three possible reasons for this relationship.

c) In which directions does industry appear to be spreading in Japan? In what ways will this affect population distribution?

2. a) Use the world thematic maps on pages A12-A14 to compare the location and other geographical characteristics of Japan with those of the United Kingdom.

b) What characteristics make Japan and the United Kingdom similar kinds of regions? What makes them different?

Further Explorations

1. Japan has a variety of traditional activities, some of which are sumo (Japanese wrestling), judo (which developed from jujutsu), and kyudo (Japanese archery).

a) Explain what is involved in one of these activities, including equipment, rules, movements, special vocabulary, and so on.

b) Write a short report on the historical development of the activity that you chose.

c) Research and report on traditional Japanese activities in this country. Your report could range from a detailed study of a local club or business to a national overview.

You may be able to demonstrate techniques for the rest of the class. Alternatively, you might make a videotape to show to the class.

2. Japan has been criticized by nations from around the world for the way in which it exploits resources and creates pollution.

a) Find and report on at least five separate pieces of evidence on this topic.

b) What efforts has Japan been making to address the problems of pollution? How successful has it been?

c) Summarize your own feelings about the criticism that Japan receives.

3. One of the factors that determines where vacation resorts will be established is the climate. Other factors include ease of access to large numbers of people, scenic beauty, and cultural attractions.

a) Imagine that it is your job to develop resorts that will increase tourism in various parts of Japan. Look at maps of Japan. Think of ways in which you would attract more tourists to each main island, and also to visit the Ryukyu Islands in the south. Make a chart with the following headings:

Island	Tourist Attraction	Additional Notes

In the Additional Notes column, you should indicate a precise location and the specific advantage that your island or location has for the tourist attraction you choose.

b) Choose one of the tourist attractions you listed in a). Explain in detail how you would develop and publicize it. Your answer should include:

- diagrams and/or maps
- a publicity poster or pamphlet
- a short written account to justify your scheme

Figure 9.18 *Japan's sandy coastal areas attract visitors.*

10

The Netherlands

A DEMOGRAPHIC REGION

Figure 10.a *Weiringermeer Polder*

Figure 10.b *House Barges on a Canal in Amsterdam*

Figure 10.c *A Street Scene in Amsterdam*

Figure 10.d *Lichtenvande*

Figure 10.e *Old Fort Heusden*

STARTING OUT

1. Describe the location of the Netherlands in relation to:
   - the rest of Europe;
   - your home.
2. What do the photos suggest about the population patterns of the Netherlands?
3. Why do you think the Netherlands are so flat?
4. Based on these photos, write five captions to attract tourists to the Netherlands.

LEARNING DESTINATIONS

At the end of this chapter, you should be able to:

- understand what constitutes a demographic region;
- explain why the Netherlands can be defined as a demographic region;
- recognize some of the unique problems faced by the people of the Netherlands;
- understand how the people have surmounted many of their problems through careful planning, wise decision making, and the use of technology;
- identify features of the Netherlands that appeal to tourists.

Introduction

In 1986, the Netherlands completed work on the Rhine Delta Project. The Project, which was first proposed in 1953, eventually cost US$5.4 billion. Over 30 km of dikes were built to block the sea from four fingers of salt water in the southwestern part of the country. The Delta Project required innovative solutions to construction problems that had never been encountered before, and took the skills of workers on a scale unheard of in public works. The completed dikes converted estuaries (inlets of the sea) into fresh-water lakes and protected the surrounding shore areas from the actions of waves and storms. It will stand as one of the largest engineering feats humankind has ever undertaken.

Why would a country of only 15 million people attempt such a huge and expensive undertaking? What do they hope to gain by taming the seas? The actions were prompted by an age-old concern for the Dutch: the need for land for the growing population. As early as the twelfth century, the people of the region were creating barriers to block the sea from their homes and fields. As the population increased, more and more of the sea had to be driven back. Each further increment of development demanded more elaborate and expensive projects.

Because land claimed from the sea in these kinds of ways is expensive, it must be used economically, particularly if much of it is designated as farmland. Residential land is, therefore, limited, and population densities are high. As a result, the Netherlands is the most densely populated country in Europe. For this reason, the country has been used to illustrate the concept of a **demographic region**. The term "demographic" refers to the population of a country, especially its size, density, and growth rates.

CHECKBACK AND APPLY

1. Most people have quite distinct images of the Netherlands. Perhaps your images include the popular ones of wooden shoes, tulips, and canals.

   a) In two minutes, write down as many images as you can of the country.

   b) Use these images either to describe an imaginary scene in the country or to sketch the scene.

2. Using the map on page 157, locate the places where the dikes of the Delta Project have been built.

   a) Why do you suppose these locations were chosen?

   b) What are the benefits of straightening the shoreline in this part of the region?

   c) Where in the country might be the next target for draining? Give reasons for your answer.

3. **a)** What other options, besides dike building and draining, does a country like the Netherlands have to deal with a growing population? Suggest at least three different options.

   b) For what reasons do you think the people chose to engage in dike building, instead of selecting one of the other options?

Figure 10.1 *A Dike Near Emmeloord*
Notice how the dike is raised well above the land. This is to hold back high water that might occur during storms.

The Netherlands as a Demographic Region

The Netherlands is located on the delta plain at the mouths of the Rhine, Maas, and Scheldt rivers. The delta formed through the accumulation of sediments dropped by the rivers over thousands of years. The plain seldom rises above 100 m above sea level, and parts of it are now as much as 6.7 m below the sea. The low elevations mean that flooding by the sea is a possibility.

About the fifth century B.C., the earliest settlers of the area built low hills onto which they could retreat during floods. By the twelfth century A.D., these hills had been linked to form dikes, and settlements sprung up in the protected places. Windmills were commonplace by the fifteenth century. These originated with the Arabs, who used windmills for irrigation.

With the windmills pumping water from the low areas, swamplands could be converted to farmlands. This led one observer to comment that "God made the Earth, but the Dutch made Holland." The newly drained **polders** were very fertile, and food production kept pace with the growing population of the region. The introduction of steam-driven pumps in the nineteenth century made the process of dike building and draining the land even more efficient.

The twentieth century has been the heyday of **land reclamation** (the reclaiming of land from the sea) in the Netherlands. The Zuider Zee was cut off from the sea by a 32 km dike by 1932 and made into a fresh-water lake. With the destructive waves of the sea gone, dike building is much easier. By the late 1980s, 1838 km² of the Zuider Zee had been changed into land for farms and cities. The Delta Project, described earlier, was just the latest phase of land reclamation in the region.

As each new parcel of land is drained, settlers arrive to make it productive. The first couple of crops are ploughed under to help improve the texture of the soil and to reduce its salinity. Within several years, the land reaches full productivity. Today, about 50 percent of the Netherlands is reclaimed land and more than one-third of the country is below sea level. Almost 60 percent of the land is used for agriculture, with farms averaging 15 ha in size. The high costs of the land, and its scarcity, mean that it must be used intensely if farmers are to realize profits. All available land is used, including land along rights of way for railways and roads.

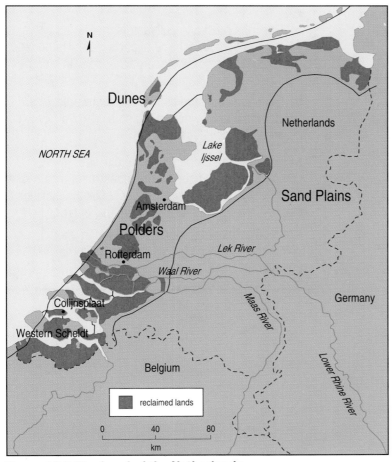

Figure 10.2 *The Land of the Netherlands*
About 50 percent of the land of the Netherlands has been reclaimed from the sea. How much more land could likely still be reclaimed?

Because of the need for good farmland, the size of urban development is restricted. The majority of the population lives in the western part of the region, which includes the cities of Amsterdam, Den Haag (the Hague), Rotterdam, and Utrecht. Amsterdam is the main commercial and cultural centre of the country; Den Haag is the capital city; Rotterdam's port handles more cargo than any other port in the world; and Utrecht is the focus of the main railway lines. The population density is particularly high in this area, and the cities have grown together to form a **conurbation**, an almost continuous urban area.

As in most countries of the world, there has been a shift in the population from rural areas to cities. This **urbanization** is a result of several factors, including the increased use of machinery in agriculture. Displaced workers have generally moved to cities to find jobs, and consequently the cities have grown rapidly.

As cities grow in population, they must expand either upwards or outwards. Because it is expensive to construct taller buildings, most cities choose to expand their boundaries to accommodate their increase in population. This puts pressure on the farmlands to change to urban uses. "Pressure" may be felt in the form of higher land and service taxes, heavier traffic on local roads, complaints from nearby urban dwellers about the sights and smells of farms, and offers to purchase land from developers. As the pressure continues, land once used for food production is converted to urban uses, such as residential areas and places of employment. Land is also used to put in place the **infrastructure** to support the growing population. The infrastructure is made up of all the facilities that must be provided before an activity can take place. In the case of urban growth, this includes roads, sewer systems, water treatment plants, municipal services (such as parks, community centres, cemeteries), and sites for the disposal of wastes.

CHECKBACK AND APPLY

4. **a)** Draw a time line to show the development of the technology used to reclaim land in the Netherlands.

    b) Speculate on what the next forms of technology for land reclamation will be like. Write a paragraph describing your ideas.

5. Locate the Netherlands on the maps on pages A2-A3 and A24-A25.

    a) Give at least three reasons to explain why Rotterdam has developed into the busiest port in the world.

    b) For what reasons is the port at Amsterdam not as important as the port at Rotterdam?

6. People's decisions to move from one place to another are influenced by push and pull factors. **Push** factors are those aspects of their present situation that they find undesirable; **pull** factors are aspects of other places that they find attractive.

    a) Describe push and pull factors that would encourage people in the rural parts of the Netherlands to move into the cities. Record your ideas in a chart format.

    b) Describe push and pull factors that would encourage a landowner on the outskirts of a city to sell property to a developer.

7. **a)** Estimate the proportion of land in your own community that is used for the infrastructure of the community. Begin by making a list of all the facilities that you know of that would fit into this category.

    b) About what proportions of the land in the community are used for residential, industrial, and commercial uses?

The Population of the Netherlands

The cities are pushing out onto the farmland because of urbanization, not because of a rapid growth in population. In fact, compared to many other countries in the world, the Netherlands has a low population growth rate. During the 1980s, the nation grew by about 0.4 percent per year. This compares to 0.8 percent for all of Europe, 1.9 percent for Asia, and 2.9 percent for Africa.

A good way to put the population growth rate into perspective is to calculate the length of time it will take for the population to double in size. The **doubling time** for the Netherlands, at its current rate of growth, is 180 years. Figure 10.3 shows doubling times for some other countries around the world.

A low population growth rate is typical of developing countries where standards of living are high. Families are usually small, with some citizens choosing not to have children at all. This situation exists because:

- parents are confident that the few children they bear will survive to become adults (infant mortality rates are low);
- people make lifestyle choices that emphasize careers, travel, and consumerism;
- the costs of raising children from birth to adulthood are prohibitively high;
- the state carries much of the burden of caring for elderly and infirm citizens;
- knowledge and use of family planning methods are commonplace.

Figure 10.3 *Doubling Time for Selected Countries*

Country	Growth Rate (%)	Doubling Time (years)
Japan	0.5	144
United States	0.6	120
USSR	0.8	90
Cuba	1.2	60
Haiti	1.9	38
Peru	2.1	34
Jamaica	2.2	33
Egypt	2.5	29
Kenya	3.9	18

In some developed countries, Canada and Denmark for instance, birth rates have fallen so low that the population is not replacing itself.

The population of the Netherlands has expanded somewhat in recent years as a result of immigration. Some of these immigrants have come from the Mediterranean area in search of better job opportunities and more prosperous lives. Other immigrants arrived from places that were part of the Dutch empire in the seventeenth century. In 1975, for example, Suriname (formerly Dutch Guiana) won its independence from the Netherlands. Thousands of people left that country just before independence to start new lives in the Netherlands.

All residents of the country are eligible for the social security benefits that are provided by the government. These benefits include unemployment insurance, sick pay, and free education. Benefits are paid out for as long as people need them and they are indexed to inflation. For the most part, the social security system is supported by the sale of natural gas that is found in the northern part of the country.

Figure 10.4 *A Street Scene in Amsterdam*
The high cost of living and the comprehensive social security system encourage people to keep their family sizes small.

CHECKBACK AND APPLY

8. Doubling times are determined using the Rule of 72 in this way:

$$\text{Doubling Time} = \frac{72}{\text{\% population increase}}$$

a) Use the rates of population increase to calculate the doubling time for Canada (0.9%), Pakistan (3.2%), Sweden (0.2%), and Mexico (2.2%).

b) Population growth can come from two sources, natural increase (more births than deaths) and net migration (more people enter the country than leave). Of the four countries listed in a), which are likely to have higher natural increase rates than net migration rates? Why?

c) In what ways is doubling time a useful indicator of population growth? What are some weaknesses of this method?

9. Compare the problems faced by countries with high rates of population growth with those with rates close to zero. Use a chart format for your answer.

10. a) What specific advantages does a country gain from having a comprehensive (thorough) social security system?

b) Where does the money to run a social security system usually come from?

c) What problems might a country like the Netherlands encounter in providing its social security system?

The Cities of the Netherlands

Since the great majority of people in the Netherlands live in cities, we shall examine them in some detail.

AMSTERDAM

Amsterdam has been called a living museum because of its historic nature and pleasant residential character. With a population of about 700 000, it is hardly a museum, but it does have a great deal of importance as an economic and cultural centre. It reached its heyday in the seventeenth century when Dutch fleets sailed from Amsterdam to exploit the riches of the West and East Indies. Wealth from these enterprises financed the construction of a grand city built on the design of expanding horseshoe canals that fit one within the other. Many mansions and public buildings remain to attest to the splendor of this time period. There are over 1000 bridges in the city, and the canals are lined with houseboats. The city is linked to the North Sea by the North Sea Canal, one of the world's deepest and widest canals.

Through immigration, Amsterdam has acquired a more diverse population. People from Suriname and Indonesia, as well as other countries, have chosen this city as their home. They have helped to give Amsterdam a cosmopolitan air that most people find charming and stimulating.

Figure 10.5 *Historic Buildings in Amsterdam*
Why might Amsterdam attract tourists to the Netherlands?

ROTTERDAM

Rotterdam is the second largest city in the Netherlands, with a population of 574 100 (1988). It lies 20 km from the mouth of the Rhine, a major river that leads directly to the heart of Europe. In the mid-1800s, Rotterdam could only be reached from the North Sea by a twisting, shallow arm of the Rhine as it spilled out across its delta. In the 1870s, a deep canal was dredged in from the sea so that larger ships could reach the port, and the city began to prosper. Now, all along the shore of the river are docks, warehouses, shipyards, oil refineries, flour mills, container terminals, storage areas for bulk cargoes, and factories. Day and night, barges and ships move through the harbour and trucks and trains rattle along the shore. Rotterdam is the busiest harbour in the world.

Much of the cargo handled at Rotterdam is not headed for the Netherlands. Instead, it is destined for some of the most populated and urbanized parts of Europe. Fanning out from Rotterdam across the Netherlands and nearby countries are roads, railways, and pipelines. River barges and inland ships can travel from here almost 800 km up the Rhine River to Switzerland. Rotterdam has aptly been called the "Gateway to Europe".

CHECKBACK AND APPLY

11. **a)** What problems might a city built on a series of canals, like Amsterdam, face?

    b) What might be some solutions to the problems?

12. In what ways would the people of the Netherlands benefit from the movement of cargo through the port of Rotterdam?

13. **a)** Examine the air photograph in Figure 10.6 and the land use map of Rotterdam in Figure 10.7. Identify each of the activities shown on the map on the air photograph. Which activity occupies the greatest amount of land?

    b) Suppose you did not have the map to show you the location of land uses. What evidence would indicate that this was a large port?

    c) What problems might this area face because of its physical environment?

Figure 10.7 *The Port of Rotterdam*

Figure 10.6 *A Land-Use Map of Rotterdam*

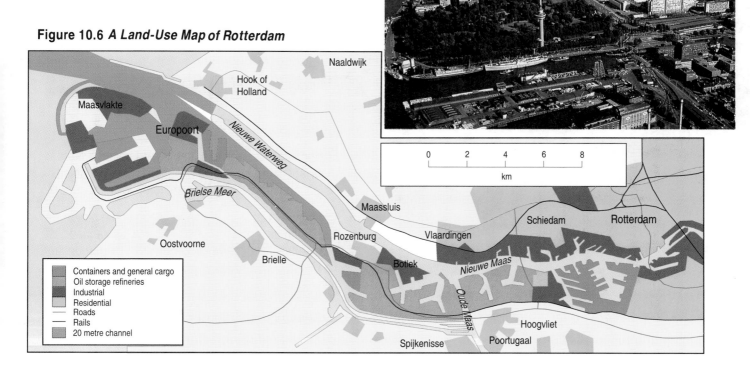

Legend:
- Containers and general cargo
- Oil storage refineries
- Industrial
- Residential
- Roads
- Rails
- 20 metre channel

The Economy of the Netherlands

How has the Netherlands managed to maintain such a high standard of living when there is so little land to meet the needs of the people? The answer lies in the wise use of natural resources, a high level of technology, successful economic strategies, and an educated population.

The major natural resources are natural gas, crude oil, agricultural land, and fish. Reserves of natural gas and oil have cushioned the effects of world energy supply problems. In addition to providing an important energy source, these reserves also provide the raw materials used in producing petrochemicals, which are the basis for a multitude of modern products, from fertilizers and chemicals to pharmaceutical goods and plastics. The agricultural land has been won through hard work, determination, and the use of specially developed technologies.

Figure 10.8 *Shipbuilding at Alblasserdam*
What locational advantages does the Netherlands have for shipbuilding?

Using the land, the Netherlands has become one of the world's largest cheese manufacturers. The North Sea and Ijsselmeer are important fishing grounds which provide employment for many people. The major fish species are herring, mackerel, cod, haddock, and eel.

The high level of literacy of the people is crucial to the development and maintenance of the healthy economy. Highly sophisticated industries depend on a well-educated population. Manufacturing employs 30 percent of the workforce. Service industries require educated people to respond to the needs of an aging population, to supply educational and medical personnel, and to design, build, and maintain the infrastructure of the nation. Two out of every three workers are employed in service industries.

Since 1948, the Netherlands has been linked to many other European nations in matters of trade. The latest phase of European interdependency, called the European Economic Community, is slated for final agreement in 1992. It provides close economic, monetary, and political ties, which should benefit all of the nations involved. It is intended that there will be a common currency, common passport, no internal tariffs, free movement within the borders for citizens of all countries involved, and no restrictions concerning work location. In other words, the distinction between countries is becoming very weak. The new Europe will be more like a "United States of Europe".

CHECKBACK AND APPLY

14. **a)** What industries have been historically important for the Netherlands?

    b) Speculate on which industries will become more important in the future. Give reasons for your choices. Which industries are likely to become less important?

15. Agriculture is important in many parts of the nation, and Dutch products are internationally acclaimed. Despite this, agriculture contributes only 4 percent to the Gross Domestic Product (GDP). With this in mind, devise two arguments for, and two arguments against, any further land reclamation.

Tourism in the Netherlands

People visit the Netherlands to enjoy the pleasures of the cities, to experience the bounty of the polders, and to marvel at the accomplishments of the Dutch people. In 1987, the country earned US$2.7 billion from tourism, a value which ranked twelfth in the world. The majority of tourists arrive from other European countries.

The relationship the Netherlands has with the sea is clearly demonstrated in the cities. The ports see huge volumes of ship and barge traffic as goods come into Europe and leave from Europe destined for the rest of the world. Because of this, the cities are lively and international in character. The mix of historic architecture and scenic canals makes walking in the cities a pleasant pastime. Roads and pathways are well lit at night. Crime rates are low. The overall atmosphere in the cities is relaxed and charming.

Tulips are a major crop in the Netherlands and an important tourist attraction. However, the dairy industry is the most important form of agriculture. It, too, attracts tourists. Much of the milk is made into cheese which is sold at the open-air market in Alkmaar. Here, traditionally dressed cheese guild workers carry platforms of cheese wheels from the warehouses to the buying areas. Tourists can purchase cheese at the local shops.

The dikes, polders, and lock systems of the country draw tourists who come to marvel at the engineering accomplishments of the Dutch. About 1000 of the traditional windmills have been preserved to remind both tourists and residents alike of the country's struggle with the sea. The dikes often serve as pathways for hikers and bicyclists, and small boats travel along the canals. During the winter, skating on the canals is a popular activity. The fresh-water lakes created by drainage schemes are destinations for tourists. Lake Ijssel (the fresh-water lake formed when the Zuider Zee was cut off from the North Sea) is a popular fishing area, and some of the surrounding villages recreate traditional ways of life for the benefit of tourists.

Figure 10.9 *Cheese Guild Workers at Alkmaar*
What advantages does the Netherlands have for the dairy and cheese industries?

CHECKBACK AND APPLY

16. What attractions might the following people wish to visit in the Netherlands?

- a history teacher
- an agricultural worker
- an engineer
- an urban planner
- a biologist
- a ship builder

17. In what ways might the growth of cities, and their expansion outward, affect the tourist industry?

18. Examine the statistical information about the Netherlands given in Figure 10.10.

Figure 10.10 *Tourist Spending and Receipts, 1987*

	Tourist Spending by Netherland Residents	Tourist Receipts in the Netherlands
Value (US$, millions)	6352	2666
Per Person in the Netherlands (US$)	434.6	182.1
Percentage of GDP	2.9	1.2
Percentage of World Tourism Spending	4.3	1.7

Source: World Tourist Organization

a) Does the Netherlands have a tourism surplus or deficit? Explain your answer.

b) What countries would you expect to be important destinations for tourists from the Netherlands? Why?

c) Suggest three actions that the tourist industry of the Netherlands might take to try to encourage more foreigners to vacation in the country.

d) What three actions could the tourist industry take to encourage more residents to vacation within the country?

The Netherlands

THE REGION AT A GLANCE

Area	40 841 km²
Population (1990)	14 765 000
Population density	361.5 persons/km²
Urban population	89%
GDP* per capita (1989)	US$13 065
Life expectancy	
—males	74 years
—females	81 years
Birth rate	13/1000
Death rate	9/1000
Population	
—under age 15	19%
—over 65	12%
Main industries	metals, machinery, oil refining, diamond cutting, food processing
Main exports	foodstuffs, machinery, chemicals, oil products, plastics
Main imports	raw materials, consumer goods, transportation equipment

*Gross Domestic Product

Famous Tourist Destinations

Tulips

The Netherlands is famous for its tulips. Tulips were first brought to the Netherlands from Turkey by Dutch traders in the seventeenth century. They quickly became popular with the Dutch and other Europeans. There are instances recorded when enormous sums of money were paid for single bulbs of particularly desirable colours. Eventually, government regulations stabilized the industry. Since then, tulips have become an important product of the region, filling a worldwide demand. More than four billion tulip bulbs are sold annually in 36 countries. At blooming time, tourists crowd the roads and nurseries of the country, particularly in the flower region west of Amsterdam, enjoying the spectacle of hectare upon hectare of brilliant colour. Flower festivals and parades are designed to attract tourists.

Flower auctions take place in Aalsmeer, near Amsterdam. As many as 2000 buyers bid on cart loads of tulips using the "Dutch auction method". A clock-like bidding wheel starts at a price higher than reasonably expected, and slowly moves to lower prices. The first person to bid pushes a button to stop the bidding wheel. The system is automated so that the purchase price is instantly withdrawn from the purchaser's account and deposited in the seller's account. Most of the flowers sold at this auction are exported.

CHECKBACK AND APPLY

19. a) In what ways is the Dutch auction method different from that used at a typical North American auction?

b) What are the advantages of the Dutch auction method?

Figure 10.11 *Tulips Near Bollenveld*
Why are these fields used for flowers instead of food? What determines the crops that are produced on agricultural land?

TOURIST GUIDE

- The Dutch strive to accommodate foreigners, especially in languages. For example, instructions in telephone booths may be printed in Dutch, French, English, and German.
- Outdoor cafes provide good opportunities to observe street life in the larger cities of the Netherlands.
- The diamond trade is a unique industry that is centred on Amsterdam. Several firms offer tours and viewing rooms where tourists can watch artisans cut, polish, and mount diamonds.
- Dutch foods are simple and hardy, and make ample use of cheese and fish.
- In the main square of Utrecht is a small statue of Anne Frank, a young German-Dutch Jewish girl who died in a concentration camp during World War II. The courage and humanity displayed in the book, *The Diary of Anne Frank*, have inspired countless people.

The Future of the Netherlands

In the Netherlands, as in all countries, the economy rises and falls with global conditions. But the Dutch economy is diverse and is able to survive poor conditions. The well-developed social security system in the Netherlands provides a buffer against hard times for citizens. Overall, the country rates very highly in terms of health and social security.

The high standard of living will encourage a lower birth rate, perhaps below the replacement level. Liberal immigration policies will continue to attract people, and the population of the country is expected to grow slowly over the coming years. This growth will no doubt continue to put pressure on the farmlands around cities, making expansion of polders necessary.

The Netherlands is destined to be an important power in Europe over the coming years.

CHECKBACK AND APPLY

20. **a)** If you were a citizen of a developing country who was seeking to emigrate, what would attract you to the Netherlands? Make a list of the country's attractive features.

b) In what ways might the country's social security system be an attraction for immigrants?

c) What sorts of demands might a high immigration rate put on the Netherlands' social security system? How should the government deal with these demands?

d) Suggest ways in which the country benefits from immigration. You should include economic, social, and political benefits in your answer.

Chapter Review

- The Netherlands is one of the most densely populated countries in the world.

- In order to obtain more land to accommodate the growing population, the people of the Netherlands have drained parts of the sea.

- Land gained through reclamation is expensive and is used sparingly, mostly for agriculture. Other uses of the land are related to shipping, such as port facilities and terminals.

- The growing urban areas are putting pressure on the surrounding farmlands, and some land has been taken over for urban uses.

- The country has a high standard of living and a low birth rate. Some of the population growth has come from immigration.

- Some tourism is focused on the cities, some on agricultural uses of the land, and some on the land reclamation projects.

Vocabulary Review

conurbation	polder
demographic region	pull factor
doubling time	push factor
infrastructure	urbanization
land reclamation	

Thinking About . . .

THE NETHERLANDS AS A DEMOGRAPHIC REGION

1. **a)** Give three reasons to explain why the people of the Netherlands have put so much effort and money into land reclamation.

   b) Has the time and effort spent on land reclamation been justified? Support your opinion with evidence.

2. Describe the locational advantages of the Netherlands for the following industries:

   - tourism
   - cheese making
   - tulip growing
   - diamond cutting
   - oil refining

3. **a)** Explain why the population density of the Netherlands is the highest in Europe.

   b) Suggest three ways of dealing with the pressures of population growth without expanding the cities onto farmlands.

   c) Draw a flow diagram to show the relationships between population growth and the loss of agricultural land.

4. Examine Figure 4.12, which shows the population of the Netherlands for the years 1829-1990.

   a) During which years did the population grow the fastest? the slowest?

   b) What events during the time period shown might have influenced the rate of population growth? What evidence suggests that the events did change the growth rate?

   c) Estimate the total population of the Netherlands for the years 2000, 2025, and 2050. Based on your estimates, will urban pressures on rural land continue to be a problem? Explain your answer.

5. What are three conditions that could hamper the growth of tourism in the Netherlands?

6. Using the map of the Netherlands on page 161, decide where in the country would be the best place to locate each of these tourist activities:

   - a beach resort/gambling casino
   - a theme park
   - a covered sports stadium

   For each activity, give at least three factors that you would consider important in making your decision.

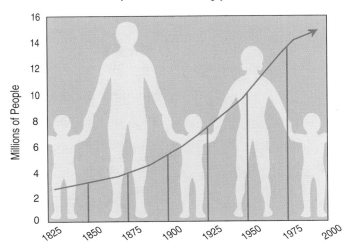

Figure 10.12 *Population Growth of the Netherlands, 1829-1990*

Atlas Activities

1. Refer to the map and Landstat image of Lake Ijssel on page A32.

   a) What is the relationship between the appearance of reclaimed areas and the time period of their reclamation? Suggest reasons for the pattern you identify.

   b) Use the satellite image to help identify the area that was reclaimed most recently. Give the name of this area. What clues led to your conclusion?

   c) Identify the location of Amsterdam on the satellite image. What evidence suggests that the waterway that connects Amsterdam to the North Sea is not a natural river?

   d) Why was the construction of the waterway from Amsterdam to the North Sea necessary?

2. Refer to the map on pages A24-A25.

   a) In what ways do land uses in the Netherlands differ from land uses in other European countries?

   b) What is the relationship between major cities and the dominant land uses around them throughout the whole of Europe? Explain why the pattern exists.

 # Making Choices

A possible land reclamation project in the Netherlands is the Wadden Shallows.

To better protect the northern coastal areas, and to increase the land area significantly, the Dutch may turn their attention to the Waddenzee. This area is shown in Figure 10.13. Much of the sea between the West Frisian Islands and the mainland is very shallow. In fact, at low tide, some of the sea floor is exposed.

Plan how best to reclaim some or all of this shallow area from the sea.

1. Enlarge the map in Figure 10.13 onto a large piece of paper. Include only the outline of the land, the scale, and north sign at this time.

2. **a)** Decide which areas you will reclaim and which you will leave as open water.

   b) Decide which of the water areas are to become fresh and which are to remain connected to the sea.

   c) Decide where you will build barrier dams and dikes around each of your polders.

   d) Design a network of main roads to link the areas of reclaimed land with each other and with existing main roads.

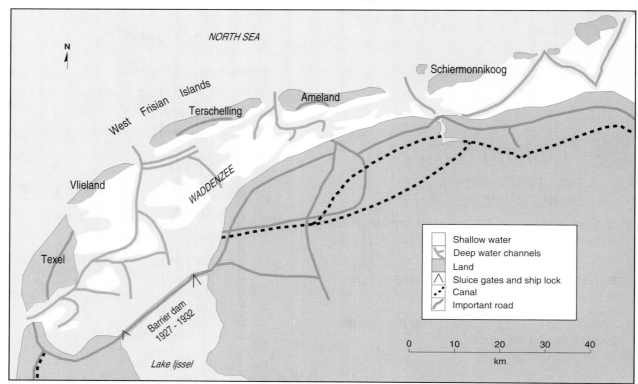

Figure 10.13 *The Depth of the Water in the Wadden Shallows*

LOOKING BACK AT THE NETHERLANDS

3. Carefully transfer your decisions to your enlarged map. Include a key and a title.

4. **a)** Determine a schedule for the project. In which order will you build barrier dams and dikes and reclaim land? Ideally, you should be able to reclaim some land as soon as possible, and the rest in stages.

   b) Assuming that you can progress at the following rates, calculate the anticipated start and finish dates for each stage of your plan. Indicate this on your map by writing the dates on the appropriate parts of the map. Alongside each stretch of barrier dam, write the dates, such as "2001 to 2003".

   Rates of Progress
   - barrier dam, 4 km per year per set of equipment
   - dike around polder, 20 km per year

   Two sets of equipment for building the barrier dam are available, and there are many sets of equipment for building the polder dikes.

   Remember: No progress can be made on polder dikes until the sea water is excluded from that area.

5. Choose one of the polder areas that you will be reclaiming. Draw an enlarged outline of the shape of the polder and the adjoining land.

6. Decide on a basic plan for the chosen polder.

   a) Indicate the location of the enclosing dike, the major canals, and any pumping stations.

   b) Show where the main town and villages will be located. Connect these with main roads.

   c) Colour in all of the land according to a suitable distribution of land uses. These might include: agricultural, residential, industrial, recreational. Use symbols to indicate where you would put buildings, such as schools and hospitals.

   d) Finish this map by including a title, key, and any explanatory notes that you think might be needed.

Further Explorations

1. Make a detailed study of the Delta Project.

   a) Draw a map of the delta area showing the structures that were built and the resulting distribution of fresh and salt water.

   b) Produce a history of the Delta Project. This can be presented in table form, listing dates and events from the plan's conception to its conclusion.

   c) Use illustrations that help to explain the new technology that was devised especially for this project. Explain why new methods were necessary.

2. Research the mining industry from mining to marketing. When you have gathered your information, produce an illustrated account that traces one jewellery-quality diamond from its discovery in a South African mine to its sale in a store in Amsterdam. Try to include some explanation of the meaning of "carat", how the quality and value of a diamond is judged, the security surrounding a diamond, and the cutting, grinding, and mounting processes.

3. Obtain tourist brochures and maps of the Netherlands by writing to the Netherlands Travel Office or by visiting a local travel agent. Use these brochures and maps, and library resources, to design a bike tour of the Netherlands. Be sure to explain:
   - your route (a map is necessary), with stops marked, a key, scale, title, and north sign;
   - what you would see, experience, and learn as you travel;
   - what you would see. experience, and learn at your stopping places.

4. Make an appointment to see the personnel manager of a motel or hotel near your community.

   a) Determine the kinds of employment opportunities that are available.

   b) Report on one kind of employment that interests you. Include information on:
   - the kinds of duties, hours, and salary involved;
   - the qualifications needed and what employers look for in a potential employee in that position;
   - the frequency of openings in that position;
   - the possibilities for advancement, and what you would need to do to make progress.

   c) Suggest ways in which the qualifications for a similar job in the Netherlands might be different.

11

Thailand

A CULTURAL REGION

Figure 11.a *Playing Music for the Dead, Bangkok Crematorium*

Figure 11.b *Workers in Rice Paddies in Southern Thailand*

Figure 11.c *Flower Market, Bangkok*

Figure 11.d *Bangkok*

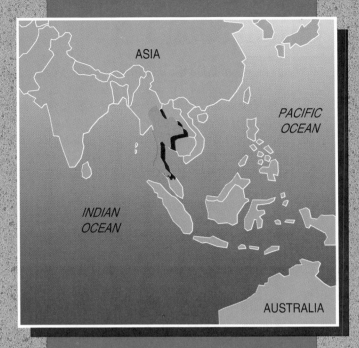

Figure 11.e *The Golden Temple, Bangkok*

GETTING STARTED

1. Study Figure 11.a. Comment on the cultural mix shown.
2. List the main elements of a culture. Include such things as language, religion, technology, customs, and dress. What cultural elements on your list are evident in the photos?
3. If you were to vacation in Thailand, what aspects of daily life in the country might be similar to those of your own life? Quite different?
4. Why might Thailand be especially attractive for visitors from North America or Europe? Give three reasons.

LEARNING DESTINATIONS

At the end of this chapter, you should be able to:

- understand what constitutes a cultural region;
- explain why Thailand may be defined as a cultural region;
- understand the historical influences on the culture of the Thai people;
- describe features of Thailand's cultural landscape;
- recognize important problems faced by the people of Thailand and identify some possible solutions.

Introduction

❝As I stepped out of the air-conditioned hotel onto the Bangkok street, I was hit by a hot, humid, wall of air, as hot as any I could remember. Just to the right of the hotel were two beautiful jacaranda trees in full bloom. Their vivid red flowers made a spectacular contrast with the grey sky. Just a few steps away from the hotel was a major street. It was jammed with trucks, cars, taxis, bicycles, pedicabs, buses, and pedestrians, all jostling for space.

I walked down the sidewalk to find out more about this delightful place. All along the sidewalk people were selling almost anything you would want to buy — from pens and pencils to freshly fried breakfast and tropical fruit. The aromas of food mixed with car exhaust fumes, the delicate scents of flowers, and the odour from the open sewers created an unforgettable blend.

I turned off the street and headed down a narrow sidestreet. To my surprise, I had stumbled upon one of the markets where flowers are sold. It was like a tropical paradise. The flowers, in every colour of the rainbow, were mostly woven into arrangements, such as wreaths and garlands. One of the vendors explained to me that people buy these flowers to take to their Buddhist temples, which can be seen almost everywhere. Some of these temples are small spirit houses that might be built in front of a home. Others are full-sized temples that take up an entire city block.

I continued on to explore new sights and sounds in this fascinating country.❞

Hugh Welbourn

This visitor to Thailand, like many others, was delighted by the charm of the people and their way of life. Thailand's culture has developed over many centuries. Although it has changed somewhat over time, it has always remained true to the fundamental principles on which it was built. One of these principles is Buddhism. The people's strong belief in Buddhism is reflected in their day-to-day activities. Because of its well-defined cultural identity, Thailand may be considered as a **cultural region**.

CHECKBACK AND APPLY

1. **a)** Identify words or phrases in the quotation that describe some cultural characteristics of Thailand.

   b) What aspects of the culture are not described in this quotation?

2. If you were to travel to Thailand, what cultural characteristics of the region would you probably find most interesting? Give reasons for your answer.

Figure 11.1 *Food Vendors in Bangkok*

The History of Thailand

People first inhabited the area that is now Thailand more than 5000 years ago. They earned their livelihood by cultivating rice. The first Thai kingdom was established in A.D. 1238. Contact with Europeans began in the 1500s, and, in the following years, the Thais traded with Portugal, Spain, England, France, the Netherlands, as well as Japan. Many European ideas were taken back to Thailand by Thai students who were educated in European schools. In spite of the contact, Thailand was never colonized by an outside power: it is the only nation in Southeast Asia that has always kept its independence.

In 1932, a rebellion in the country forced the reigning king to change the government from an absolute monarchy to a **constitutional monarchy**. The monarch now shares power with an elected government. In spite of changes in the government, some of them quite abrupt, the royal family provides a stability that helps to unite the factions within the country.

The name of the country was changed from Siam to Thailand in 1939. Two years later, Japan invaded Thailand, and, in a move to protect its own security, the Thai government signed a treaty of alliance with Japan. At the conclusion of World War II, Thailand developed stronger ties to the United States. The country managed to avoid being drawn into the conflicts between Communist forces and groups that sought to block their expansion that rocked Southeast Asia during the 1960s and 70s. Over the last decades, it has prospered.

Because of the country's history, the Thai people have remained homogeneous: more than 85 percent of the population shares a common culture and speaks the same dialect. The largest minority group is Chinese, who make up about 12 percent of the population. The Chinese have become well integrated into Thai society. In addition, immigration quotas now restrict the number of Chinese immigrants to Thailand; consequently, the Chinese influence on the Thai culture has been limited.

CHECKBACK AND APPLY

3. The Thai people are proud of the fact that they were never colonized by a European country.

    a) In what ways might colonization have changed the country?

    b) Suggest some ways in which Thailand may be different from its neighbours which were colonized by Europeans.

4. Thailand is a homogeneous nation because the vast majority of its people share a common heritage. Other countries have heterogeneous populations; that is, they contain many ethnic groups and have no single common heritage. Canada has a heterogeneous population since its largest ethnic group (British) makes up only about 40 percent of the population.

    a) Identify ways in which homogeneous populations will be different from heterogeneous ones.

    b) What are three advantages and three disadvantages of living in a homogeneous population? of living in a heterogeneous population?

Figure 11.2 *Ploughing a Rice Paddy in Rural Thailand*
Cultivation of rice has been an important part of the culture of the region for thousands of years.

Thailand as a Cultural Region

The king is required by the constitution to be a practising Buddhist, and Buddhism is the official state religion. It is the most visible force that binds the people of the country, with about 95 percent of the population claiming adherence. Some of the main principles of this religion involve doing good to others and harming no living thing. A person's position in society and well-being is, according to Buddhism, a direct result of his or her behaviour in previous lives. Throughout the land, temples — *wats* — dot the landscape. Over 40 percent of males serve in the Buddhist monkhood for at least a few months; in fact, the number of monks has prompted some to refer to the country as "the land of the yellow robes". Popular religious beliefs and institutions have remained almost unchanged over the centuries.

Thailand is a rural nation. More than 60 percent of the citizens make their living by farming. Some 80 percent of the population live in rural villages that range in size from several hundred to a few thousand people. The villages fall into three types:

- cluster settlements, in river valleys;
- strip villages, along rivers or canals;
- dispersed settlements, on river deltas.

The population density is highest in the agricultural areas of Thailand's central lowlands, where it averages about 250 persons/km². Each village boasts a school and a *wat*, which form the social centre of the community.

The population shift from rural areas to cities, a common occurrence in many other parts of the world, has not significantly affected the

Figure 11.3 *A Gold Statue of Buddha*
This 5 t statue is located in one of Bangkok's many temples. Has Buddhism become a part of Canada's multicultural heritage?

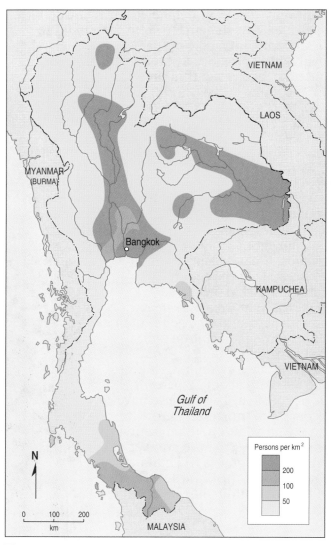

Figure 11.4 *Population Density of Thailand*

Figure 11.5 *A Village in Rural Thailand*
Why do you suppose it was necessary for the farmers to
create terraces in this hillside?

people of Thailand. About nine out of every ten
people live and die within a few kilometres of
where they were born.

In most of rural Thailand, the extended
family is the basis on which life operates. People
live in housing compounds along with their
relatives, including uncles, aunts, cousins, and
grandparents. Tradition is strong, and the way
of life is based on the interactions of people at
the village level.

Thai houses, usually made of wood and
thatch, are built on stilts along canals and
rivers. The stilts provide protection from the
floods that occur during the monsoon season,
between July and December. In cities, stucco
houses are more common.

Women and men have equal rights in
Thailand, and women are well represented in the
labour force. However, until recently, there were
few women in professional or governmental
roles. This is changing, and women now sit in
the national legislature. During elections,
women participate as fully as men in the
political process.

Recognizing the needs of a rapidly growing
population, the government of Thailand has a
population policy that is aimed at reducing
growth rates. Family planning programs
encourage the use of contraceptives. Growth
rates in the country have fallen from 3.1 percent
in 1980 to 1.3 percent in 1989.

CHECKBACK AND APPLY

5. **a)** What factors might encourage Thais to remain in
their villages, participating in a traditional way of life?

   b) What forces might draw the people towards the
cities?

6. **a)** Thailand has a relatively young population, with
roughly half of its people under the age of 20. What
difficulties might Thailand face in meeting the needs of
these young people?

   b) In the future, as the young people grow up, what
problems might the government face?

7. Imagine two countries, one in which women have
political and legal rights and the other in which they do
not. Compare these two countries using a chart format.
Some characteristics you may wish to compare are:

- role of women in the work force
- influence of women in political life
- standards of living for women and children
- size of families
- stability of family units

Discuss your points with others to discover their
opinions.

Thailand

THE REGION AT A GLANCE

Area	513 999 km²
Population (1990)	54 890 000
Population density	106 persons/km²
Urban population	20%
GDP* per capita (1987)	US$840
Life expectancy	
—males	62 years
—females	68 years
Literacy	89%
Work force	
—agriculture	59%
—industry, commerce, services	26%
Arable land	34%
Main exports	textiles, rice, fish products, rubber, tin
Main imports	petroleum, machinery, food, fertilizer, chemicals

*Gross Domestic Product

Famous Tourist Destination

Bangkok

No other city in Thailand comes close to Bangkok's size or importance. With about 6 million people, it is many times larger than the second largest city in the country; over one-half of the nation's urban dwellers live in or near Bangkok. This domination, both in size and importance, makes Bangkok a **primate city** — the most important city in the country. It is expected to reach a population of 7.6 million by the year 2000.

The hustle and bustle of this large and fascinating city draw tourists. The city has a rich mixture of old and new, rich and poor, quaint and exciting. Vendors and hawkers line the streets, while temples and shrines provide oases of quiet and solitude. New factories and apartment buildings are popping up like mushrooms around the city, standing shoulder to shoulder with ancient buildings. These exciting contrasts lure vacationers to the city.

However, with size and growth come problems. In Bangkok, the number of cars and other vehicles is increasing so rapidly that the city streets are often choked with traffic and air pollution levels are high. Tourists have no difficulty finding taxis or other vehicles in which to travel around the city, but the traffic jams make for slow trips.

Figure 11.6 *Scenes From Bangkok*
These two scenes suggest some of the contrasts that are found in Bangkok. The three-wheeled taxi cabs shown in the photo on the right are called *tuk tuks*.

Another of Bangkok's problems is also a result of its growth. As the number of residents and visitors increases, so does the demand for water, and the city's wells are pumped at faster rates than they can handle. The result is that the city, which is built on soft **alluvial deposits** (deposits of sand or mud), is sinking at a rate of about five to ten centimetres a year. This has led to increased flooding from the oceans as well as damage to important buildings. Only a major effort to slow pumping of water, and to develop alternative sources of water, will avert serious trouble.

About one-third of Bangkok's residents are Chinese in origin. They dominate finance and industry in the city and run 80 percent of the rice trade, including exporting, wholesaling, and retailing. The Chinese people, because of their differences in backgrounds, are not a unified group and have largely been integrated into Thai society.

CHECKBACK AND APPLY

8. If you were to visit Bangkok, what features of the city would you find interesting? Give reasons for your choices.

9. **a)** In what ways might the values and attitudes of the residents of Bangkok differ from those of the people of rural Thailand?

   b) What strains or problems might this pose for the country?

   c) What do you suppose are some of the factors that bind the urban and rural people of Thailand together?

TOURIST GUIDE

- Thailand has an extremely strict anti-narcotics law. Severe sentences, including the death penalty, are given for using and trafficking in drugs.
- Be sure to show proper respect for all images of Buddha or the royal family.
- Customs regulations prohibit the export of religious images, no matter how old, without permission of the government.
- The head is considered to be the most respected part of the body. It is an insult to touch or pat anyone on the head, with the exception of little children. On the other hand, the feet are the least respected part of the body, and it is considered an insult to point your feet at someone.
- No immunizations are required for entry into Thailand. However, travellers to rural areas are advised to have shots for typhoid, tetanus, rabies, and malaria.
- Comfortable, lightweight, washable clothing is most practical for Thailand's tropical climate. Average temperatures reach 37°C in May in Bangkok.

Tourism in Thailand

Tourism generates about two billion dollars annually for the country, 4.1 percent of its GDP. Much of the tourist activity is centred on the culture of the people. Visitors have opportunities to view many architecturally and historically unique buildings, ruins, and monuments. Among the major attractions are the Grand Palace and the Reclining Buddha. In the northern highland area, visitors encounter tribes that maintain a way of life that has varied little over the centuries.

In southern Thailand, many modern beach resorts have been developed, modelled on European or American resorts. These meet the needs of tourists who are seeking comfortable surroundings, fine cuisine, and a night life. Many well-known hotel chains have facilities in Thailand.

The development of these resort facilities has attracted other enterprises that service tourists. For example, some familiar North American franchises, such as Pizza Hut and Dunkin' Donuts, now have outlets in Thailand. Fast food is not, however, new to the country: there are thousands of street vendors who cook food for passersby. Adventurous tourists can snack on paper-thin dried squid, deep-fried shrimp, or *satay* — meat served on bamboo skewers. Other services available to tourists include buses and taxis. Some of the modern intercity buses used by tourists show videos and have attendants.

Not all tourists, however, use modern means of transportation while in Thailand. Some people prefer to experience the unusual, as the following account testifies.

> I had never ridden an elephant before and was a little apprehensive about what I was getting into. But, at the appointed time, we met the man who was to take us by elephant deep into the jungle. He was very friendly and obviously knew his elephant well. We climbed from a small platform about three metres off the ground onto a wooden saddle on the back of the elephant, and off we went.
>
> At first the path was fairly wide, but, as we went higher and deeper into the jungle, it narrowed and the hillside dropped sharply off to the right-hand side. The path was muddy and slippery, but the guide reassured us that the elephant used his trunk to keep from falling. I still didn't like to look down to the river far below.
>
> As we rode, we jostled back and forth. The saddle was not tied as securely to the elephant as a horse's saddle would be, and we felt that we could slip off its back. Our guide sat right up on the head of the elephant. When we went through thick brush, the elephant would fold his ears back over the legs of the guide to keep him from falling off. The elephant and his owner were very close. We had to watch for tree branches that the elephant would brush aside but which would swing back and hit us.
>
> It seemed that the elephant never stopped eating. All the way up the trail, the elephant used his trunk to pull over branches and leaves to munch on. When he came to a stream, he would take a drink and then use his trunk to splash water on his belly and back. Needless to say, he splashed the riders on his back as well.
>
> We were in a totally different world from the busy streets of Bangkok. The jungle was lush and green and full of animal life. When our ride ended, the elephant went to haul logs in the forest as he had for many years. Unfortunately, the tropical forest which is his home is disappearing quickly and with it his future and that of many elephants like him.

Lily Marcinek

Figure 11.7 *A Market in the Chinese District of Bangkok*
Why might tourists find this setting attractive?

CHECKBACK AND APPLY

10. Suppose you were a resident of a village in Thailand. What might be your reaction to the tourists who come to your village and visit your temple?

11. a) For what reasons would international hotel chains build facilities in a country like Thailand?

    b) If you were to travel to Thailand, would you prefer to stay in an internationally owned hotel or in accommodations run by Thais? Explain your choice.

12. In what ways would income earned by elephant owners benefit many people in a village? Use the idea of the ripple effect (see page 17) in your answer.

Figure 11.8 *Elephants Cooling in a River*
Elephants like these are used both as workers and in the tourist industry.

Thailand's Future

Thailand is one of the most prosperous nations in Southeast Asia. It has abundant natural resources and an increasingly well-educated population. Because of its political stability, the country has been able to take actions against such problems as uncontrolled population growth and a dependence on primary industries. Its future is bright.

However, there are clouds on the horizon. One problem is the large number of refugees who live within the borders of the country. Thailand served as the **country of first asylum** (the first place for refuge) for hundreds of thousands of people fleeing the turmoil of Kampuchea, Laos, and Vietnam in the mid-to-late 1970s and the 1980s. Although the government gave safe haven to these refugees, it has not offered them resident status. United Nations' assistance maintains the refugees until such time as they can be resettled in another country, or until they can return to their homelands. As long as they stay in Thailand, the refugees need food and other assistance and, consequently, put a strain on Thailand's resources.

Another set of problems arises from the rapid industrialization that has taken place in the country. Manufacturing (of cement, food products, wood and paper, and textiles) has become increasingly important in Thailand. Large multinational companies have set up branch plants to assemble electronic equipment, automobiles, and other products. These companies are attracted to Thailand by the absence of strict air and water pollution regulations. Not having to meet tough control measures means that costs can be kept low. As a result, air quality has declined rapidly in the past few years and now threatens the health of millions of people.

Most of the industrial activity in the country is concentrated in the Bangkok area. The rest of the country, however, remains relatively poor. In these areas, health care is inadequate, schools are undeveloped, all-weather roads are lacking, and safe water is not available for many of the people.

Another problem is linked to the international drug trade. Northern Thailand includes an isolated mountain area, known as the Golden Triangle. Mountain farmers in this area grow poppies on a large scale. When harvested, these poppies form the basis of the international heroin trade. Although these poppies provide an income for the farmers of the region, the drug trade presents serious problems for Thailand. The crime and corruption that go along with drug smuggling have brought misery to many in Thailand, and drug dependency is a growing problem.

Thailand is located in an area of the world that has seen much political strife and war over the past decades. The country must be in a constant state of alert to defend its boundaries and protect its citizens from the turmoil around it. The occasional conflicts between Thailand and Kampuchea pose real threats that could erupt into international war at any time. Yet, in the midst of this uncertainty, Thailand continues to experience real development and growth.

CHECKBACK AND APPLY

13. **a)** From your reading, list the forces or issues from the world outside Thailand that are affecting Thailand.

    b) Suggest what effects each would have.

    c) Explain how changes in transportation and communication in the world are influencing Thailand.

14. Review the problems described in this section of the chapter. Decide which of these problems will be most important for:
    - average citizens of Thailand
    - the government of the country
    - tourists in the country

    Give reasons to explain your decisions.

15. As with all countries, the government of Thailand has limited financial and human resources and so must make choices in dealing with problems. If you were in a position of authority in the government of the country, which problem would you tackle first? (In part, your decision might be based on the chances of achieving success in solving the problem.) Describe some actions you would take to solve the problem.

Figure 11.9 *Urban Poverty and Wealth in Bangkok*
Poverty and wealth exist side-by-side in urban areas of Thailand, as they do in the large cities of the rest of the world.

Chapter Review

- Thailand's culture has developed over many hundreds of years.

- Since the country remained independent and was not colonized during the period of European expansion into the area, its culture grew and changed without foreign influence.

- Religion forms an important part of the Thai culture, and temples and shrines are characteristic of the cultural landscape.

- The Thai people have maintained strong ties to the land and agriculture. The great majority of people continue to live in rural areas.

- Bangkok is the largest city and the centre of industrial activity in Thailand.

- Tourism is often based on the culture of the people and the distinctive landscape of the region. In recent years, the beaches have played a larger role in the tourist industry.

- Many of the problems the country faces result from circumstances that originate outside of Thailand, such as military aggressiveness and the demand for drugs.

 # Vocabulary Review

alluvial deposits
constitutional monarchy
country of first asylum
cultural region
primate city

 # Thinking About . . .

THAILAND AS A CULTURAL REGION

1. a) Explain why Thailand can properly be considered as a cultural region.

   b) Which aspects of the Thai culture are most similar to your own culture? Which are most different?

   c) Is Canada likely to develop a homogeneous culture (see page 138)? Explain your answer.

2. Describe the locational advantages of Thailand for:
   - rice farming
   - tourism
   - manufacturing

3. a) Identify the major changes that are taking place in Thailand and describe how they are affecting the country.

   b) What specific characteristics of the Thai people and the Thai culture have allowed these changes to occur?

   c) What concerns do you have about the future of Thailand? Explain your answer.

4. a) Identify three conditions that could hamper the development of the tourist industry in Thailand. At least one of your conditions should relate to Thailand's place in the international scene.

   b) What three conditions could improve the development of the tourist industry in the country? Explain your points clearly.

5. Examine Figure 11.10 showing the populations of both Thailand and Bangkok.

   a) About what proportion was Bangkok's population of all of Thailand's in 1980?

   b) How is this proportion changing?

   c) Calculate the rate of growth for Thailand from 1980 to 1990. Use the following formula:

$$\frac{\text{population for 1990} - \text{population for 1980}}{\text{population for 1980}} \times 100$$

Calculate the projected rate of growth for 1990 to 2000 using the same method.

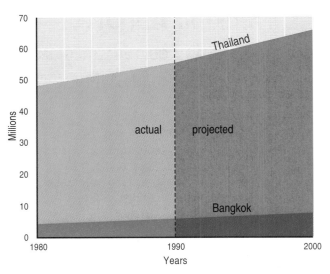

Figure 11.10 *Population of Thailand and Bangkok, 1980-2000*

Atlas Activities

1. Turn to pages A2-A3 in the atlas section of this book.

   a) Describe Thailand's location in the world with respect to important bodies of water, including rivers.

   b) Describe Thailand's location in Asia with respect to large and important countries.

   c) Identify advantages and disadvantages of Thailand's location for its tourist industry. You might use a chart to record your ideas.

2. Examine the map on page A4 and describe the pattern of population density in Southeast Asia. List and explain the major factors that account for this pattern.

3. The maps on page A14 show the patterns of major religious groups and languages.

   a) What percentage of the world's people are Buddhist?

   b) What parts of the world have large numbers of people who adhere to Buddhism?

   c) Compare the distribution of Buddhists in the world to other major religious groups. What conclusions can you draw?

   d) What language groups are found in Thailand and Southeast Asia? What conditions would account for their distribution in the area? Explain.

 # Making Choices

The government of Thailand has established a task force to study conditions in the country and to propose improvements that could be made. The task force has four members; each person is specially trained to examine one area in depth and make concrete proposals that will be presented for consideration and possible funding. The proposals are to be designed in such a way as to produce the greatest benefit in the least complicated way, and with the least expense.

The expertise and specific jobs are outlined below.

a) *Tourism Specialist:* This individual has a wide range of experience in the travel and tourism industry throughout Asia.
Task: to identify areas or sites in the country that have potential for tourism, and then to develop them as fully as possible. One approach would be to list the types of facilities that tourists need and the changes that must be made to accommodate such facilities.

b) *Small Business Expert:* This person has extensive experience in helping people set up small businesses. Small businesses are the backbone of the Thai economy, and include shops, small manufacturing plants that employ under ten people, and services such as haircutting, polishing shoes, and transporting goods.
Task: to make recommendations that will spur the development of small businesses of all kinds. Some questions this person will consider include: What do small businesses need to start up? How can the government help? What services are essential for a small business?

c) *Forestry Expert:* This forester has worked for years to get logging companies and farmers to use good management practices in cutting rainforests.
Task: to develop a plan to slow the destruction of the forest and to promote the replanting of trees. Some questions for consideration are: What groups of people play important roles in using the rainforests? What could be done to save the rainforests, but still provide an income for the people who live in the hill country?

d) *Child Welfare Expert:* This individual has a large job ahead — to identify possible methods of improving the health, diet, education, and general welfare of the children of Thailand. Many of the children suffer from malnutrition and its related diseases, such as rickets and scurvy, as well as other contagious diseases, including measles, mumps, and tuberculosis.
Task: to come up with recommendations that will help to improve the quality of life for the children in the country. It is essential that this person develop a list of groups that must be consulted on this issue, as well as an action plan to address the most important concerns immediately.

1. In groups of four, decide which member of the task force each person will be. Each person is to prepare a report that follows this format:

   • a list and brief description of key issues

   • a list of the groups of people who must be consulted while preparing this report

   • possible solutions for each of the issues listed

   • a description of work that must be done in the future after this report is adopted

   • obstacles that must be faced in order for this report to be implemented

 Further Explorations

1. Select a country that borders on or is close to Thailand, such as Laos, Vietnam, Malaysia, or Myanmar.

   a) Describe briefly the political history of that country over the last 30 years.

   b) Where is this country located geographically? In addition to giving its general location in southern Asia, list the countries that surround it.

   c) What specific geographic features are found in the country?

   d) What signs are there that this country is either developing economically or having economic troubles?

   e) What problems does this country face at the present time?

   f) What tourist potential does this country have? Give details: describe possible tourist sites and give the current number of tourists or value of income from tourists.

2. Research the history of Thailand.

   a) List some of the key events that have taken place in Thailand over the last 700 years or so. Emphasize the unique or unusual developments that have helped to mould Thailand into the country that it is today.

   b) Outline five distinctive aspects of Thai culture that have not been discussed in this text.

   c) Change is occurring very quickly in Thailand, and this change is altering almost every aspect of life in the country. From your research, describe as many ways as possible in which change is affecting life in the country.

12

Siberia

A FRONTIER REGION

Figure 12.a   *The Village of Kurdyum, Altai Territory*

Figure 12.b   *A Lake Near Norilsk*

Figure 12.c   *Autumn Colours of the Chukotka Tundra*

Figure 12.d   *The Altai Taiga*

Figure 12.e   *A Herd of Deer, West Siberian Plain*

STARTING OUT

1. Describe the characteristics of the natural environment of Siberia (including landforms, climate, and vegetation) that are shown in the photos.
2. What aspects of the built environment suggest that this is a frontier region?
3. What characteristics of Siberia might appeal to tourists?
4. Based on your observation of the physical and built environment of Siberia, what major problems do you suppose this region is facing?

LEARNING DESTINATIONS

At the end of this chapter, you should be able to:
- identify the characteristics of a frontier region;
- give evidence to show that Siberia is a frontier region;
- understand how careful planning and the application of technology have helped to solve some of the problems in Siberia;
- identify features of Siberia that might appeal to tourists.

Introduction

Siberia! Few places on Earth have such a universally bad reputation as does Siberia. The very name conjures up images of barren, frigid lands; of workers forced to labour under desperate conditions; and of living standards barely above those of some of the poorest countries in the world. To some extent, the reputation is deserved: ruthless Soviet leaders did send political prisoners to Siberia to work in labour camps. But the popular image of Siberia does not reveal the whole picture. Under closer scrutiny, Siberia can be seen as a land rich in resources and populated by ambitious and willing people. It is a land whose natural and human riches have just begun to be tapped.

The Soviet Union is determined that Siberia's resources will be unlocked and made available to the people and industries of the country. Because Siberia is such an immense northern land, the task seems overwhelming. Once resources are discovered, their extraction has to be planned, transportation routes constructed, cities and factories built, workers recruited. The central government of the Soviet Union plays a role in this by co-ordinating these various activities and by supplying construction materials. Imaginative solutions to truly unique problems have been, and continue to be, found.

As each new project is undertaken, the land becomes more developed. The roads built to link new communities become the supply routes for the next project. The process is on-going; each year, more and more is known about the region and more resources are extracted. This push to develop a previously unknown and unexploited area is the reason why Siberia may be defined as a **frontier region**. Its full potential has yet to be determined.

CHECKBACK AND APPLY

1. **a)** In your own words, explain what a frontier region is.

   b) Name two or three frontier regions that exist in other continents.

2. **a)** List some natural resources that you would expect to be found in Siberia. Record your ideas under these headings:

   - minerals
   - energy
   - flora
   - fauna

   b) Suggest which resources are likely to be most in demand by Soviet industries. Give reasons for your answer.

3. **a)** For what reasons might tourists be interested in visiting a frontier region?

   b) If you were to travel to Siberia, what are three topics that you would like to learn more about?

Figure 12.1 *A Typical Landscape in Siberia*
What are the opportunities of this land? What are the challenges?

Siberia's Physical Environment

Siberia, with an area of 10 968 600 km², makes up about half of the area of the Union of Soviet Socialist Republics (USSR). It sweeps from the Ural Mountains, which mark the boundary between Europe and Asia, all the way to the Pacific Ocean and the Bering Strait, a total distance of approximately 6000 km.

Siberia has three main physical areas — the West Siberian Plain, the Central Siberian Plateau, and the Eastern Siberian Mountains. These areas are marked on the map in Figure 12.3. Note how the elevations of the land affect the directions of the the main rivers in Siberia. The mountain chains were thrust up during periods of mountain building throughout the Earth's history. The fold mountains in the east are still quite active and form part of the Pacific's Ring of Fire.

Figure 12.2 *Areas of Selected Places (km²)*

USSR	22 402 100
Siberia	10 968 600
North America	24 350 000
Canada	9 970 600
United States	9 372 600
Asia	45 100 000
China	9 596 900
South America	17 870 000
Brazil	8 511 900
Europe	9 840 000
Africa	30 097 000

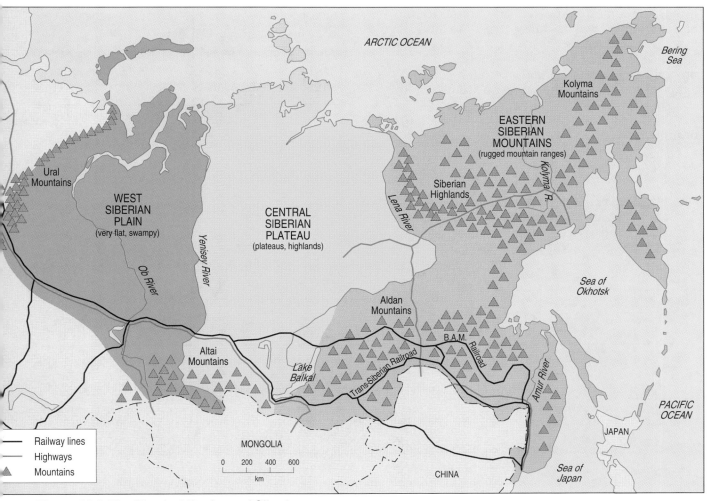

Figure 12.3 *The Physical Regions of Siberia*
Siberia is slightly larger in area than Canada. What other similarities exist between the two places?

Siberia has a **continental climate**, with low precipitation levels and wide temperature ranges. Oceans exert little influence on Siberia's climate: the Atlantic Ocean is too far away; the Arctic Ocean is frozen for much of the year; and the eastern mountains block the flow of air from the Pacific area. As a result, precipitation totals are less than 500 mm. Temperatures in the northeast range from 14°C in July to -50°C in January, while in the southwest a typical temperature range would be from 18°C to -22°C.

Permafrost — a layer of permanently frozen ground — covers most of the eastern part of the region. Only the top few centimetres of the soil thaw during the summer. Since water cannot seep through the permafrost, it sits near the surface, forming swamps.

The most northern part of the region has a **tundra** (treeless plain) ecosystem in which mosses and lichens dominate. Trees do not grow in the extreme cold. This zone extends several hundred kilometres south of the Arctic Ocean, and then gives way to hardy trees, Most of Siberia is covered by the **taiga**, a vast coniferous forest. Only in the southwest, where precipitation is somewhat lower and evaporation rates higher, does the forest yield to the grasslands of the **steppe**.

Wildlife in the region includes reindeer, fox, ermine, lynx, and sable.

CHECKBACK AND APPLY

4. What problems does the large size and physical environment of the region pose for the development of resources?

5. The people of Siberia refer to permafrost as "eternal frost". In what ways might permafrost both help and hinder development of the region?

6. Examine the climate statistics in Figure 12.4.

   a) Calculate the temperature range for each of the three locations. (Temperature range is the number of degrees between the highest and lowest monthly temperatures.)

   b) Name two factors that will influence the temperature range of a location.

   c) Compare the amount of precipitation that the three locations receive to the amount of precipitation in your area. In what ways would you expect the vegetation of Siberia to be different from that in your area?

   d) An old proverb says: "Winter in Siberia lasts twelve months. Summer lasts for the rest of the year." Is this an accurate statement about the climate of Siberia?

Figure 12.4 *Climate Statistics for Omsk, Okhotsk, and Verkhoyansk*
The top row for each place gives the average monthly temperature in degrees Celsius. The bottom row gives average monthly precipitation in millimetres.

Cities	J	F	M	A	M	J	J	A	S	O	N	D
Omsk (55°N, 73°E)												
Temp. (°C)	-22	-19	-13	-1	10	16	18	16	10	1	-11	-18
Precip. (mm)	15	8	8	13	30	51	51	51	28	25	18	20
Okhotsk (59°C, 143°E)												
Temp. (°C)	-24	-22	-14	-6	2	7	13	13	8	-3	-15	-22
Precip. (mm)	3	3	3	5	13	28	13	46	53	18	5	5
Verkhoyansk (68°N, 134°E)												
Temp. (°C)	-50	-45	-32	-15	0	12	14	9	2	-15	-38	-48
Precip. (mm)	5	5	3	5	8	23	28	25	13	8	8	5

The Resources of Siberia

Eighty-five percent of the population of the Soviet Union lives west of the Ural Mountains, and 75 percent of the industrial activity occurs in that area. It is the industrial heartland of the country, but an area rapidly running out of natural raw materials. The industries need the natural resources of Siberia, such commodities as minerals, forest products, energy, food, and furs. The problem is that these resources are thousands of kilometres away from factories and cities. Some industries are being relocated in Siberia in order to be closer to the raw materials. More often, however, the focus is on getting the resources to the industries as efficiently as possible.

In the Soviet Union, the state owns all the natural resources. It is believed that this is the best way to control the resources so that the people of the country receive the greatest benefit. The state co-ordinates the activities that are necessary to develop the resource industries properly.

There are many minerals in the folded rocks of the Central Siberian Plateau and the Eastern Siberian Mountains. Gold, platinum, tin, iron ore, copper, tungsten, and diamonds are mined in remote and isolated locations. There are also mineral fuels — coal, oil, and natural gas. There may yet be more minerals: geological surveys are still incomplete. Finding these resources is only one part of the battle; extracting them and making them available to consumers is a larger problem.

Siberia's transportation system is poorly developed. There are few roads, and many of those that do exist are gravel-surface roads which are adversely affected by the weather. Only one major railroad line crosses the region, the Trans-Siberian Railway. The largest cities of the region are concentrated along this line or its branches. Long-distance airlifts for freight are prohibitively expensive. The rivers of the region have little potential for boat transportation because they flow from south to north and are frozen for much of the year.

Figure 12.5 *Coal Mining in Krasnoyarsk Territory*
This photograph shows the equipment used to mine for coal in Siberia.

However, the rivers can be used to transport softwood timber. Logs from the great northern forests are floated down rivers, such as the Yenisei, to ports on the Arctic Ocean. Here, they are loaded onto ships for export to the western USSR or Japan. Ice breakers keep a sea route open for about four months each year. The most important trees are spruce, pine, fir, and larch. These species are suitable for the construction and furniture industries, pulp and paper, and as cellulose for plastics. Some wood is processed in Siberia using the hydro-electric power produced by power stations on the rivers of the region.

The taiga is also home to another resource, fur-bearing animals. Professional hunters and trappers harvest the fur crop for sale to citizens of the USSR and for export. Reindeer herding takes place in the far north of the region.

CHECKBACK AND APPLY

7. a) Identify several important problems that limit the development of resources in Siberia.

b) For each problem you identified in a), describe several possible courses of action to deal with the situation. In your answer, discuss who should be responsible for solving the problem.

8. Why do most industries remain in the western part of the country, rather than move closer to their raw materials? Give three reasons.

9. Suppose you were made responsible for developing Siberia's transportation systems.

a) What forms of transportation would you choose to emphasize in the region? Why?

b) Which forms of transportation would seem to be the least practical for carrying goods? for carrying passengers? Explain your answers.

c) Make a sketch map to show the transportation systems you would develop in Siberia. Write a short paragraph explaining the advantages of your system.

The Role of the Central Government

Under the Soviet constitution, all means of production, including the land, are publicly owned. The central planning agency decides how the country's resources are to be used and co-ordinates all aspects of their development, including such things as the quantities that are to be produced, how products will be distributed, labour requirements, and energy needs. Details are worked out in Five-Year Plans that set goals for the country as a whole. Since planning of this type began in the 1920s, large projects involving heavy industries have taken precedence over the production of consumer goods.

The centrally planned nature of the Soviet economy means that resource development projects do not have to be justified solely on the grounds of supply and demand. In this command economy, projects are undertaken because they achieve national goals. One of these goals over the past decades has been the industrial development of Siberia.

Large-scale development of the region started during World War II, when whole factories were relocated to Siberia to protect them from the Nazis. After the war, it was decided to leave the plants permanently in Siberia and to speed industrial development of the region. This would accomplish two main objectives: it would reduce transportation costs by moving factories closer to raw materials and energy sources, and it would strategically disperse economic activity. Millions of workers have migrated to new factories, mines, and plants in Siberia.

Energy is the cornerstone of development in the region. Once a plentiful supply of electric power is available, further industrialization and urbanization are possible. In Siberia, the large rivers have been harnessed to supply hydro-electric power. They now supply about 20 percent of the nation's total electrical needs. One of the world's largest power dams, the Sayano-Shushensk Dam on the Yenisei River, generates 6.5 million kilowatts. A large part of the region's energy needs are, however, met by fossil fuels. Vast oil fields along the Ob River yield over half of the Soviet Union's petroleum, and natural gas is supplied from reserves near the Arctic coast. There are also large supplies of coal. The rapid development of these resources was prompted by goals established by the central planning agency, and was possible because the necessary human, economic, and physical resources were made available.

In recent years, resources have been made available to develop the country's transportation facilities. In spite of the fact that it is the largest country in the world, the Soviet Union has less than half the distance of railroad lines as the United States does. (This is partly because the population of the Soviet Union is more concentrated than that in the United States.) Several new lines are under construction, including the Baikal-Amur Mainline. This 3200 km project will provide a second connection between the heart of Siberia and the Pacific Ocean, facilitating the export of resources, particularly to Japan. Plans are afoot to develop new industrial centres along this rail line.

All the recent developments in Siberia have created a labour shortage. Workers are enticed to the new industrial communities by incentive programs that offer higher pay and extra fringe benefits, such as longer vacation periods. Although many workers stay for only a few years, some choose to establish permanent roots in Siberia. The new communities are planned to have all the facilities that are expected in Soviet society — such as community centres, health care services, schools, and recreation opportunities. Since they are new, housing units are frequently better than the people would expect in the western part of the country. As in all societies, people make trade-offs when they choose where to live and work.

Figure 12.6 *The Sayano-Shushensk Hydro-Electric Power Station*
Why does an undertaking of this size require government involvement?

Figure 12.7 *A Residential Area in Novosibirsk*
Notice that the buildings in this community are quite new.

CHECKBACK AND APPLY

10. a) Compare decision making in a centrally planned economy to decision making in a market-based economy. Use a chart to organize your ideas.

b) What are the advantages of using a centrally planned approach to developing natural resources?

c) Suggest some disadvantages of a centrally planned economy.

11. Explain why the energy and transportation infrastructure must be in place before large-scale development of resources can take place.

12. Soviet literature calls Siberia the "giant construction site of the USSR". Many of the country's new developments are concentrated in this region. In your opinion, is it advisable for the Soviet Union to focus so much activity in Siberia? Explain.

Siberia

THE REGION AT A GLANCE

Area	10 968 600 km²
Population (1990)	33 183 000
Population density	3.0 persons/km²
Urban population	70%
GDP* per capita (USSR)	US$9211
Economic growth rate (USSR)	1.4%
Largest cities	
—Novosibirsk	1 350 000
—Sverdlovsk	1 220 000
—Chelyabinsk	1 050 000
—Omsk	1 020 000
Main industries	mining, forestry, machinery, chemicals, building materials

*Gross Domestic Product

Tourism in Siberia

All frontier regions are characterized by the fact that their potential has not yet been realized. The wealth of the mineral, forest, and energy resources of Siberia is largely undeveloped. Tourism suffers from the same unrealized development. The question for tourism in Siberia is not so much what is available, but what is possible. To put the lack of development of tourism into perspective, it is worth noting that, in 1987, the Soviet Union had tourist receipts of only US$198 000 000. This represents 0.10 percent of world tourism receipts. Most of this revenue was earned in the Moscow region and the western part of the country.

In general, the tourist-related resources of the region could be categorized into aspects of:

- the physical environment
- the culture and history of the native people
- the built environment
- scientific achievements

THE PHYSICAL ENVIRONMENT

The vast area of Siberia has such diverse landscapes as rugged mountains, flat plains, Arctic tundra, and dense forests. Tourists could enjoy high-intensity experiences, white-water rafting or mountain climbing for example, or more relaxed pastimes, such as birding or wilderness camping. Jobs could be created as hunting guides, wilderness outfitters, bush pilots, ski resort operators, park rangers, camp ground developers, and so on.

THE CULTURE AND HISTORY OF THE NATIVE PEOPLE

Half of the people of the Soviet Union are Russians, and the state has encouraged assimilation of ethnic groups into the mainstream population. Nevertheless, there are pockets of native populations in Siberia. The Yakuts were traditionally nomadic, hunting and gathering in the taiga and tundra regions. Their culture remains distinctive and interesting to those comparing historic and contemporary peoples.

Figure 12.8 *The Altai Mountains of Eastern Siberia*
Based on your knowledge of tourism in other parts of the world, what potential for the tourist industry does this place have?

THE BUILT ENVIRONMENT

The massive engineering achievements in Siberia will generate tourism. The dams, mines, cities, and factories of the region stand as tribute to the abilities of the Soviet people, as do the transportation facilities that have been carved from the wilderness. The cities often include facilities for theatres and dance companies, popular forms of entertainment in Siberia.

SCIENTIFIC ACHIEVEMENTS

The Siberians have spent considerable time and effort learning how to get the maximum benefit from the land. They lead the world in research on such topics as permafrost, reindeer herding, and construction techniques under Arctic conditions. Interest in these developments will draw travellers from other parts of the Soviet Union and from other northern countries.

The development of fast and reliable transportation linkages is critical to the growth of tourism in Siberia. The government is making efforts to improve rail and road networks — not for tourism, but for the exploration of natural resources — and the industry should expand over the coming decades.

CHECKBACK AND APPLY

13. **a)** In what ways would tourism contribute to the development of Siberia?

    b) What factors might prevent tourism from being considered in official plans?

14. Where in Siberia would be the best locations for the following tourist facilities?

    - a ski resort serving mostly foreign tourists
    - a conference centre for scientists
    - a wildlife reserve, with lodging facilities

    For each facility, list the three most important factors in determining location. Make a sketch map to show where you would locate your tourist facilities.

Famous Tourist Destinations

Lake Baikal

Thirty million years ago, intense geological forces within the Earth's crust caused the continent to crack. Part of the crust between two parallel faults collapsed to form a rift valley more than 1600 km long, 45 to 60 km wide, and as much as 5 km deep.

Lake Baikal is the world's deepest lake. It holds more fresh water than the five Great Lakes combined. Because it is isolated from other large bodies of water, it has developed its own distinctive biologic communities. Over 1000 plant and animal species, including fresh-water seals, are found only in Lake Baikal.

When European explorers and fur traders discovered the lake, they marvelled at the abundance of the fish and the animal life along its shores. Unfortunately, a number of species were almost wiped out by the greed of the newcomers. In recent years, pollution from a pulp and paper mill has damaged the lake and upset the delicate ecological balance.

Government officials have planned programs to improve the environment around Lake Baikal, but economic concerns usually take priority over protecting the natural environment

CHECKBACK AND APPLY

15. Identify the location of Lake Baikal on the map on page 187. Suppose you were a Siberian put in charge of planning tourist facilities for this region, what would you want? Write a one-page summary of your plan in which you:

   • describe the facilities you would construct;
   • note the infrastructure improvements that would be necessary;
   • identify problems that would likely be encountered in constructing the facilities;
   • describe the benefits for the area and the country.

Figure 12.9 *Lake Baikal*

The Future of Siberia

The development of Siberia has only just begun. It will be generations before the extent of the wealth is even estimated. But there are problems with which the region must deal. Pollution is one of the most pressing.

Pressure to develop the resources of Siberia as quickly as possible has meant that little time and energy were available for protecting the environment. Environmentalists point out that there is hardly one lake in the whole region that is not affected in some way. Pollution may come from chemicals or minerals from industrial activity, or from human and animal wastes. Unfortunately, planners often have to make choices between economic development and protection of the environment, a difficult decision when citizens are criticizing the government for lack of progress in improving standards of living and supplying more consumer goods.

Another long-term problem is political instability within the Soviet Union. Developing large projects is time consuming and uses huge amounts of labour and equipment. In the latter part of the 1980s and early 1990s, the nation was under a good deal of tension because of conflicting political ideologies, economic reforms, and ethnic tensions. Together, these factors have the potential to divert commitment and energies away from major resource development schemes in Siberia. The leadership of the country will have to continue to keep economic initiatives moving forward, as well as solve the nation's internal disputes.

One of the most frustrating problems facing Siberia, as well as the rest of the Soviet Union, is underperformance. This is the condition in which goals are not achieved, productivity is poor, and quality of output is substandard. Critics of the economic system argue that underperformance is a result of centralized planning. Managers show little initiative in making decisions, and workers have little commitment to the system's goals. The end result is an economy that is sluggish and unwilling to take risks, such as experimenting with new technologies. Underperformance can only be alleviated by decentralization of responsibilities and worker incentives.

The serious problems in Siberia cast a shadow of doubt over the future of the region. However, this vast land has so mush potential that it is difficult not to be optimistic.

CHECKBACK AND APPLY

16. For each of the three problems identified in this section of the chapter, note the nature of the problem, the source of the problem, its likely effects, and some possible solutions. Record your ideas in a chart.

17. In what ways might the problems identified in this section help or hurt the tourist industry?

18. To what extent does Canada suffer from the same problems as the Soviet Union? Support your opinion with examples and facts.

TOURIST GUIDE

- Tourists do not visit Siberia for fun or to relax; they do so to experience a different way of life.
- Group tours are the best way to travel in the Soviet Union, as they take priority over individuals. Stick to the itinerary of the tour.
- Travel between late spring and early fall. In winter, the climate is too harsh.
- There are no authorized auto routes for foreigners in Siberia. Trains and airplanes are the only realistic ways to travel long distances.
- Bone carvings, jasper and amber ornaments, and furs are the best buys in Siberia.
- Because of local shortages of common items, it is a good idea to take with you anything you might need. Such things as toothpaste, aspirins, toilet paper, and sink stoppers may not be available once in Siberia.
- Tipping with cash is not done in Siberia. Take along a supply of small gifts, such as paperback books, cosmetics, pantyhose, and pens, to show your appreciation for good service.

Chapter Review

- Siberia is an example of a frontier region because of its isolation and the incomplete development of the infrastructure necessary to extract the resources of the region.
- The resources of the region are vast; extracting them requires great engineering feats.
- The centrally planned nature of the Soviet economy allows massive resources to be used to reach goals.
- Each new venture opens up Siberia a little more. New towns are being built to house the people who migrate to this region.
- The tourist industry is not large yet, but the potential exists for a strong industry.
- Economic, social, and environmental problems must be solved before Siberia can reach its full potential.

Vocabulary Review

continental climate   steppe
frontier region   taiga
permafrost   tundra

Thinking About ...

SIBERIA AS A FRONTIER REGION

1. a) Identify the characteristics that make Siberia a frontier region.

   b) Review the types of regions used to focus the case studies in this book. As what other type(s) of region might Siberia be defined?

2. Imagine you are a construction worker in the Soviet Union. List some advantages of living and working in Siberia. What would be some disadvantages?

3. A Soviet official was quoted as saying: "Overall total planning is the only way to *develop* the North as opposed to simple *exploitation* of its resources."

   a) Distinguish between the terms "develop" and "exploit" as they are used in this quotation.

   b) In your own words, explain what the official is saying.

   c) Do you agree with this person's opinion? Give reasons for your answer.

   d) Would you describe the use of resources by the tourist industry as "development" or "exploitation"? Explain.

4. Examine Figure 12.10 which shows the urban and rural populations of the Soviet Union.

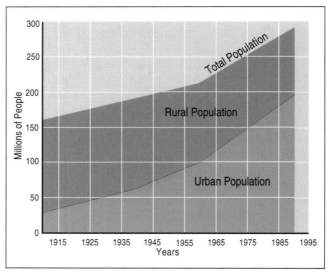

Figure 12.10 *Urban and Rural Populations of the Soviet Union, 1913-1990*

   a) Approximately what percentage of the total population was the urban population in 1920? What was the percentage in 1990?

   b) What three factors might explain this rate of urbanization of the population?

   c) This graph shows the population for all of the Soviet Union. Would you expect Siberia to show the same patterns? Explain your answer.

5. Develop an advertising campaign to attract North American tourists to Siberia. Follow these steps:

a) Identify the intended audience for your advertisements (for example, young people, affluent people, professional people, students, people of Russian extraction).

b) List the features of Siberia that you intend to promote to your audience.

c) Decide which media you will use to communicate to your intended audience.

d) Write the script for a one-minute TV or radio ad, or design a billboard poster.

How might you measure the effectiveness of your advertising campaign?

 ## Atlas Activities

1. Identify the location of Siberia on the map on page A18-A19.

a) Describe the shape of this region. In what ways might the region's shape make it difficult to develop its natural resources?

b) Identify the physical features that act as borders to the region. In what ways might these border features make it difficult to develop the natural resources?

c) List the mineral resources that are most commonly mined in Siberia. Describe the distribution of the mineral locations in the region.

2. Is Siberia most appropriately categorized as an Asian region? Give evidence to support your opinion on this topic using the pages of the atlas section, particularly pages A1, A14, and A18-A19.

3. Use a comparison chart to analyze the climates of Siberia and Canada. Data for your chart can be found on pages A11 and A12. Headings for your chart might include:

- January Temperatures
- July Temperatures
- Precipitation
- Climate Influences
- Climate Classification

 ## Further Explorations

1. Select two major problems or issues facing Siberia in the 1990s and beyond. For each problem or issue, explain:

a) its nature

b) the part of the region most affected

c) who/what it affects most

d) possible solutions

2. Compare the Siberia frontier region to any other type of region. Some headings you might use are:

- Type of Region
- Size of Region
- Characteristics
- Advantages for Tourism
- Problems

A comparison chart would be a good way to organize the information.

3. Plan a two-week visit to Siberia. Use travel brochures and tourist guides to get details about places. Include at least five cities in your itinerary, and draw a detailed map. Make reference to such practical matters as costs, travel dates, time zones, and the like. Make suggestions for clothing and other equipment that you will need.

13

Yukon
A FRONTIER REGION

Figure 13.a *Kluane National Park*

Figure 13.b *Five Finger Rapids*

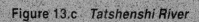

Figure 13.c *Tatshenshi River*

Figure 13.d *Ketza*

Figure 13.e *Whitehorse*

STARTING OUT

1. Describe the location of the Yukon in relation to:
   - the rest of Canada;
   - your home;
   - the frontier region of Siberia.
2. What evidence in the photos suggests that this area is still being developed?
3. In what ways are the landforms of the Yukon similar to and different from the landforms in your area?
4. Write a paragraph describing the harsh characteristics of the Yukon in a way that makes them appeal to tourists.

LEARNING DESTINATIONS

At the end of this chapter, you should be able to:
- identify the characteristics of a frontier region;
- give evidence to show that the Yukon is a frontier region;
- appreciate the importance of resource development to the region;
- recognize the constraints and opportunities of the Yukon;
- evaluate the Yukon's potential for tourism.

Introduction

❝*The prospectors came first in twos and threes with little more than a rucksack, a gold pan, a short-stemmed shovel, and a phial of mercury, living on beans and tea and bacon — men fleeing ahead of civilization. Whenever they struck it rich a circus parade of camp-followers crowded in upon them, saloon-keepers and hurdy-gurdy girls, tinhorn gamblers and three-card monte men, road agents, prostitutes, vigilantes, and tenderfeet. Sylvan valleys became industrial bees' nests, meadows were transformed into brawling shack towns; the sighing of the wind and the roaring of the river were drowned by the tuneless scraping of dance-hall violins and the crash of butchered timber, until it came time to move on to the next divide and to seek new valleys beyond unnamed mountains. And so, like the forward patrols of a mighty army, the first prospectors reached the last frontier and began, in the seventies and eighties, to infiltrate the Yukon Valley.*❞

From *Klondike: The Last Great Gold Rush, 1896-1899*, by Pierre Berton

The Yukon today is no longer the scene of gold rush activity. The gold ran out very quickly, and the prospectors and their followers headed for new places to strike it rich, leaving behind their equipment and structures, and a devastated environment. But the gold rush made Canadians, and others, aware of the Yukon as a **frontier region**.

A frontier region is one where there has been little development, and where exploration and settlement are still taking place. A frontier, in this context, is the zone in which determining the extent and value of natural resources, constructing transportation routes, and establishing functional, permanent settlements are the major activities.

Figure 13.1 *The Klondike During the Gold Rush*
Scenes like this were common during the years of the gold rush.

The impetus for development in the Yukon has come, for the most part, from three directions:

- the rising demand for natural resources, particularly minerals, to support modern, affluent ways of life;
- the United States' need to establish road links between the 48 contiguous (adjoining) states and Alaska;
- the need for improved land routes before the petroleum and natural gas deposits along the Arctic coast and the Beaufort Sea can be extracted.

Developments in the Yukon have already affected the environment and the native peoples to a considerable degree. This chapter will explore the Yukon to understand its character and issues related to its development.

CHECKBACK AND APPLY

1. **a)** In your own words, explain what is meant by the term "frontier region".

   b) Suggest some circumstances that would entice people to move to the Yukon from more southern locations. What sacrifices and risks might be involved in such a move?

2. Reread Pierre Berton's description of the Yukon (page 200).

   a) Use a dictionary and other books to find the meanings of any words or phrases with which you are not familiar.

   b) Imagine yourself as a native person who has lived in the area for many years, following a traditional way of life. Write a 200-word statement expressing your thoughts about the development of the Yukon.

3. As you work through this chapter, note ways in which the Yukon has changed since the days described in the excerpt from Berton's book. Note also the ways in which the territory is much the same as a century ago.

4. Using the atlas section of this book for reference, identify three other areas of the world which could be described as frontier regions. In what ways are these regions similar to, and different from, the Yukon frontier region? Use sketch maps to illustrate your answer.

Early Development in the Yukon

The lifestyle of the Indians of the Yukon, before European contact, involved a seasonal round of hunting, fishing, digging roots, and gathering berries. Housing was constructed of logs roofed with bark, materials found in abundance in the boreal forest, which covers most of the region. Tools, weapons, and clothing were also made from locally available resources, such as stones and animal skins. The people lived in small, closely related groups. A number of Inuit people occupied the coast of the Arctic Ocean. The native people lived in this region for thousands of years before the newcomers arrived. Fur traders were the first Europeans to make contact.

Gold! The cry went up in 1896, and tens of thousands of people headed for the Yukon in one of the biggest gold rushes the world has ever known. Unable to resist the lure of fabulous wealth, these adventurers travelled along a dozen different routes to reach the remote gold fields. Many crossed the coastal mountains by way of the rugged Chilkoot and White Passes, which led from the Pacific to the interior.

By 1899, Dawson City, capital of the Yukon and centre of the Klondike gold rush, was the home of 30 000 people and was the biggest city west of Winnipeg. There was plenty of night life, with the Palace Grand Theatre, the Eldorado Hotel, Diamond Tooth Gertie's gambling casino, a bar on every corner, and "whiskey flowing faster than the Yukon River". Many residents of the town became merchants, serving the needs of those who arrived in time to stake a claim.

Imagine the impact of the prospectors and newcomers on the environment and on the Indians of the region. Thousands of outsiders recklessly hunted game animals for food. Their carelessness caused forest fires and the destruction of animal habitats. This disruption meant that traditional Indian life was difficult to maintain. For a few years, many Indians were employed in the mines and settlements, but their employment suddenly stopped when the gold was exhausted. The Indian people returned to hunting and trapping in an environment that was badly damaged by the gold seekers.

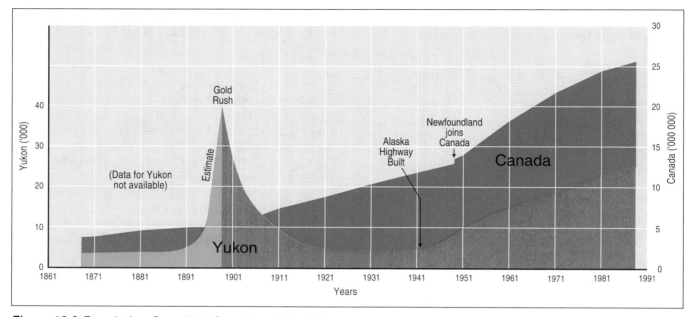

Figure 13.2 *Population Growth in Canada and the Yukon*
Read the left-hand scale for the Yukon and the right-hand scale for the rest of Canada. Describe the difference in population growth in the Yukon and in the rest of Canada. Compare the population of the modern Yukon with that of the Yukon during the years of the gold rush.

Gold valued at more than $95 million was mined from the Klondike region between 1896 and 1903, but the easily extracted gold was soon exhausted. By 1921, the population of the entire Yukon had dropped to below 5000, and Dawson City today has only 1800 inhabitants. Although the **boom** had become a **bust**, the federal government had taken the opportunity to set up a government, establish police garrisons, and had begun to provide some social services in the Yukon. In addition, a railway had been built to connect Whitehorse, Yukon, with Skagway, Alaska, on the Pacific coast, and paddlewheelers connected Whitehorse, Dawson City, and Mayo.

In the years that followed the gold rush, coal, silver, and lead mines opened in the Yukon to take advantage of rich deposits of these minerals. During nine frantic months of 1942, the Alaska Highway was built by Canadian and American workers in response to a perceived Japanese threat to Alaska. This highway crosses the southern part of the region and runs through Whitehorse. These developments encouraged more people to make the Yukon their permanent home. In 1953, the capital was moved from Dawson to Whitehorse, which had become the centre of economic activity. Today, Whitehorse has about 15 000 of the total Yukon population of approximately 26 000 people.

Boom and bust periods are common in frontier regions. The rapid growth and development that come with exploitation of natural resources often cannot be sustained over long periods.

Boom periods are triggered by:

- new discoveries of resources;
- changes in technologies that lead known resources to be more highly valued;
- construction of transportation facilities that allow resources to be exploited;
- rapidly growing demand for resources.

Bust times may be triggered by:

- depletion of resources;
- decreasing demand for resources;
- development of richer or more accessible deposits of similar resources;
- changes in technology that allow other materials to be substituted for a resource.

In the Yukon, the discovery of gold led to the boom of the gold rush, and the depletion of the mineral resulted in a bust.

CHECKBACK AND APPLY

5. In Canada, the frontier of development is progressing slowly towards the north.

a) What is the reason for this south to north expansion of development?

b) Under what circumstances might the frontier of development continue into northern locations more remote than the Yukon?

c) On an outline map of North America, use arrows and dates to illustrate the direction in which the frontier region has progressed across the land to its present location. Use an atlas and other maps of North America and Canada as reference sources.

6. a) Explain why frontier regions are unable to weather poor economic times as easily as more developed areas, such as southern British Columbia, Alberta, Ontario, or Quebec.

b) What measures might be undertaken to end boom and bust cycles in the Yukon?

c) Explain how tourism might play a role in reducing the effects of bust times.

7. a) List ways in which the Indians of the region suffered as a result of the development of the Yukon. Make a second list of ways in which they benefited as a result of the development of the Yukon.

b) Compare your lists with those of your classmates, and disuss the reasons for any differences of opinion.

The Yukon

THE REGION AT A GLANCE

Area	483 450 km²
Population (1990)	258 000
Population density	0.05 persons/km²
Urban population	64.6%
Per capita GDP* (1988)	Cdn$27 188
Change GDP* (1987-88)	10.5%
Birth rate (1988)	20.1 per 1000 people
—for Canada	14.5 per 1000 people
Death rate (1988)	5.4 per 1000 people
—for Canada	7.3 per 1000 people
Main industries	mining, tourism
Population of Whitehorse	15 199

*Gross Domestic Product

Figure 13.3 *Emerald Lake, South of Whitehorse*
What economic opportunities does this area offer?

The Yukon Today

The economy of the Yukon today depends on a few main industries — mining, tourism, timber, trapping, agriculture, and fishing. Mining continues to be the dominant economic activity, earning about $500 million annually. Gold, zinc, and lead are the leading minerals. Since these products are susceptible to world fluctuations in supply and demand, the boom and bust cycle is still experienced, although the effects are cushioned somewhat by other activities, which include government services and tourism. Forest production is limited by the small size, slow growth, and inaccessibility of timber. Most of that which is produced is for local use. About 800 trappers are registered, and the harvest of pelts is valued at over one million dollars per year. Agriculture, mainly the production of forage for animals, is increasing in importance, and the government is encouraging people to develop the industry so that more foods are raised locally. Fishing produces food for local consumption, and one processing plant in Dawson City cans salmon and caviar for export. The value of fishing activities is about $4 million each year. Tourism is an important industry, earning $37 million annually from about 500 000 visitors. It is second only to mining in earnings. Figure 13.4 summarizes employment by sector in the Yukon.

The introduction of hunting and fishing regulations and falling fur prices have made it increasingly difficult for the Indians of the region to make a living in traditional ways. As a result, the Indian people negotiated with federal and territorial governments for control of land, and for compensation of land that they can no longer use. In 1988, a land claims settlement awarded the Indians $200 million and title to 26 000 km² of land. Native people today are employed in a variety of ways: in resource development and transportation, in service industries, and in the production and sale of handicrafts.

Figure 13.4 *Employment by Sector, the Yukon and Canada, 1989*

Sector	Number Employed in the Yukon	Percent Employed in Canada
Agriculture, Fishing, Logging	70	4.4
Mining	939	1.4
Manufacturing and Construction	1 130	23.1
Transportation and Communication	1 279	7.7
Wholesale and Retail Trade	1 885	17.5
Finance and Real Estate	456	5.9
Business, Educational, Accommodation, Health, and Social Services	2 402	33.2
Government	4 346	6.8
TOTAL	12 507	100.0

Source: Statistics Canada and Yukon Bureau of Statistics

Robert Service wrote many poems about the Yukon, especially about the gold rush. His famous poem, "The Cremation of Sam McGee", tells of the bitter cold that the prospectors experienced as they travelled in the wintertime:

On a Christmas Day we were mushing our
  way over the Dawson Trail.
Talk of your cold! through the parka's fold
  it stabbed like a driven nail.
If our eyes we'd close, then the lashes
  froze till sometimes we couldn't see;
It wasn't much fun, but the only one
  to whimper was Sam McGee.

Figure 13.5 *Climate Statistics for Whitehorse*

Whitehorse (61°N, 135°W)	J	F	M	A	M	J	J	A	S	O	N	D
Temp. (°C)	-18.9	-13.2	-7.7	-0.1	7.1	12.4	14.1	12.3	7.8	0.7	-9.0	-15.8
Precip. (mm)	18	14	15	11	14	29	33	36	29	20	22	20

Climate affects development in two important ways: costs and limitations on activities. The cost of heating, which is primarily fuelled by oil, is a major expense for residents and businesses. In those northern communities built on permafrost, special construction techniques are employed to prevent the thawing of the frozen ground. Buildings are constructed so that cold air can flow underneath the floor, between the ground surface and the building. If this were not done, the buildings would slowly melt the permafrost below and would sink into the ground.

Outdoor activities are curtailed to a certain extent in the long, dark, winter months, though many mines still function under artificial light, and people enjoy typical winter sports of snowmobiling and hunting.

In the spring season, insects may limit outdoor activities, especially in wetter areas, and the short summers severely restrict agricultural activities. Animals are raised for milk and meat, but are housed for long periods in barns where they are fed on locally produced forage crops. Some short-season vegetable crops are grown but, for the most part, vegetables and

fruits are imported from the south, together with a great variety of other foodstuffs that cannot be produced locally.

Other important limits on development are imposed by the rugged terrain of the Yukon and its distance from the major population centres of North America. Almost all of the region is within the Western Cordillera landform region. This mountainous region, which extends the length of North America, is composed of high, sharp-peaked mountains and deep valleys. Rivers like the Nahanni, the Yukon, and the Porcupine tumble and churn as they find their way to the oceans. Construction of transportation routes is difficult and expensive in this landscape.

The high costs of transporting products into and out of this region prohibit manufacturing. It is simply far too great a distance to potential consumers in southern Canada or the United States to make manufacturing viable, and the population of the Yukon is too small to support industries on its own. With the exception of minerals, whose value is high enough to absorb the expense of transportation, few products are exported.

Figure 13.6 *Salmon Drying at Old Crow*
The native peoples have used the resources of the Yukon for as long as 30 000 years.

CHECKBACK AND APPLY

8. a) Using the statistics from Figure 13.4, construct **pictographs** that compare the structures of the labour force in the Yukon and in Canada as a whole. A pictograph is a graph that uses symbols instead of colours or shading. (For example, one person symbol could represent 50 people.) Be as imaginative as you can. *Hint:* Convert the figures for the Yukon into percentages. Fully label your pictographs and give them suitable titles.

b) Compare and contrast the sectors in which people are employed in the Yukon and in Canada as a whole. Suggest reasons to explain the differences by relating them to the Yukon's geographical location, natural resource base, and its comparatively young stage of economic development.

c) Suggest ways in which each sector of the Yukon's economy benefits from tourist activities. Which sector benefits the most?

9. a) Find climatic statistics for a centre near to your area. These should be available in an atlas or in a book on Canadian geography. Construct climate graphs using the local statistics and those of Whitehorse.

b) Compare and contrast temperature conditions in the winter and the summer. Refer to the maxima, minima, and range values in your summary.

c) Compare the precipitation patterns, totals, and forms of precipitation in the two stations.

d) Briefly explain the reasons for the differences or similarities that you observed in b) and c).

e) Describe the problems and opportunities the climate of Whitehorse has for tourism. Record your ideas in an Advantages/Disadvantages chart.

10. a) Using Figure 13.7, describe the **situation** of Whitehorse. A description of situation involves a concise statement explaining where a place is with respect to: political boundaries, major rivers, highways and railways, and other major communities to which it is linked.

b) Draw a simple sketch map that illustrates the situation of Whitehorse.

11. a) Using the topographic map of Whitehorse in Figure 13.8, describe the **site** of Whitehorse. A description of site is much more detailed than that of situation. It includes the dimensions of the urban area, and where it is located with respect to rivers, roads and/or railways, whether the land is flat, gently sloping or steep, the width of the river, which way the river flows and how it is crossed, and a description of the nature of the land immediately surrounding the community.

b) Draw a land-use map of the area shown in Figure 13.8. Include these uses:

- residential
- transportation and communications (roads, railways, airports, communication towers, etc.)
- institutional (schools, hospitals, etc.)
- industrial (including mines)
- commercial (businesses, malls, etc.)
- utilities (water pipelines, hydro lines, etc.)
- recreational
- forest

Identify major areas by name or symbol where appropriate. You might also consider marking significant buildings.

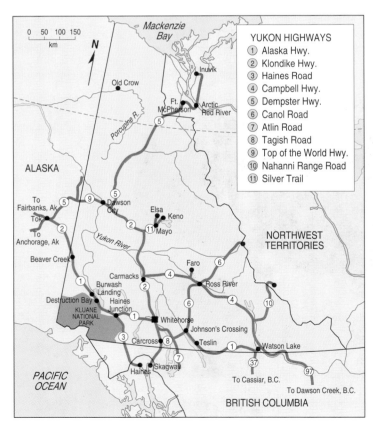

Figure 13.7 *Main Roads in the Yukon Territory*
Describe the pattern of roads in this region.

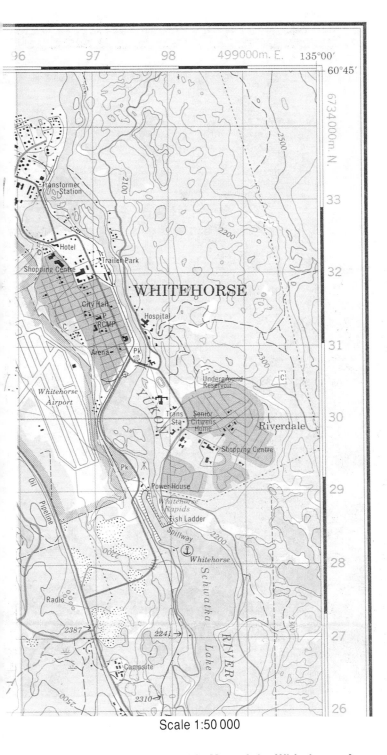

Scale 1:50 000

Figure 13.8 *A Topographic Map of the Whitehorse Area*

TOURIST GUIDE

- Many people travel to the Yukon by road. The Alaska Highway is paved and travel is not very difficult, but most other highways have loose surfaces. Tires are often damaged and flying gravel can crack windscreens and chip paint. An additional spare tire is advisable. Gas stations are sometimes far apart, so carry spare gas.
- Hiking and camping expeditions in high and remote areas must be planned carefully. In Kluane National Park, any overnight trips must be registered with park staff, and permission must be obtained for mountain climbing.
- In the summer you may be annoyed by blackflies and mosquitoes, so protective clothing and repellants are recommended.
- Grizzly and black bears live in the area. At night, use ropes to hang your food between high branches of adjacent trees; never keep food where you are sleeping. Make some noise as you travel through wooded areas so that you do not startle a bear. Never approach a bear cub or get between a mother and her cub. If you do get into trouble, climb a tree.
- Outfitters will provide a variety of services to canoers who want to get off the beaten track. Guides can be arranged, or you may prefer drop-off and pick-up services. Remote canoeing should only be attempted by experienced trippers.

Famous Tourist Destinations
Hiking the Chilkoot Trail

"We had heard about the Chilkoot Pass and the prospectors during the gold rush. Surely the Pass couldn't be as bad as the stories led you to believe! The trail is only 53 km long, and we have hiked three times that distance before. Thousands of prospectors made it across with heavier loads than we would need. After all, back in those days, the Canadian Mounties didn't allow anyone to cross without provisions for one year. Imagine carrying over 900 kg up the treacherous steps carved in the ice of the steep mountainside!

So we made our plans, and that's when we started to have doubts. Experts warned of the possibility of blinding snow, ice, sleet, fog, and gale-force winds, even in the summer. A lightweight tent, sleeping bag, and camp stove were vital, we were told. We'd also need wool and waterproof clothing and dried food. Articles in hiking magazines recommended a walking stick to help keep your balance, crampons (spikes which you attach to your hiking boots to help on the icy patches), gaiters, and zippered waterproof lower leg protectors to help keep you dry. In the end, we decided to go for it.

We arrived in Skagway, Alaska, the scene of much activity during gold rush days. It was a sunny July afternoon, so we spent the rest of the day exploring the town, with its reminders of times past.

In the morning, we rose early and packed our gear. We were lucky to get a boat ride to Dyea, 14 km away, where the trail starts. The day was fresh, and the sun had already been up for several hours when we started. The trail was marked clearly, and we eagerly pushed ahead. Along the way we spotted a bald eagle, several ravens, and lots of smaller birds. Towards evening, we pitched our tent at one of the ten campsites along the trail. We were tired, and our feet benefited from a paddle in a

Figure 13.9 *Hikers in Rough Country*
What tourist facilities would hikers like these use in the Yukon?

nearby freezing-cold stream. After a hearty meal, we packed up our food, slung it from a rope between two trees, and prepared for a night's rest. As we were drifting off, a distant rustle brought us bolt upright. A bear? But then came the babble of human voices. Relief! We peered out at six weary travellers who were also making their night's camp at the campsite. They were nearly at the end of their journey, having started at the eastern end of the trail. It had taken them three days, and the weather had been fine. They were certainly tired, though, and were looking forward to a bath in the lodge in Skagway.

The toughest part of our trip was the 45 degree climb to the top of the Chilkoot Pass. It felt like clambering up a vertical wall. Every footstep was a challenge, and the straps of the packs dug deeply into our shoulders. There was no shelter from the strong wind, for we were above the tree line. Thankfully, we made it through the pass and established our camp in a sheltered site. We were exhausted, but made the effort to change from our clothes which were wet with perspiration into dry ones. It was cold and we didn't want to get **hypothermia**, which occurs when a person's internal temperature drops too low. This can lead to death if not treated promptly, and around here...

On the last day of our hike, it was windy, cold, and raining. We were greeted by two Canadian wardens who had started out from Bennett, B.C. They were checking to see if hikers along the trail were all right in the unexpectedly cold weather. We were pleased to see them, and it was good to know that we were only six hours from Bennett.

In Bennett, we rested a while by the lake, and then hitched a ride on a boat which had just brought in some hikers who were going to travel west along the trail to Skagway. The boat took us about 40 km to Carcross, where we were going to catch a bus to Dawson City. We felt good, having finished our challenge, but for the prospectors during the gold rush, this was only the beginning: they still had a long journey down the Yukon River. "

C. Hannell

CHECKBACK AND APPLY

12. a) Using Figure 13.7, determine how far prospectors had to travel from Carcross to Dawson City.

b) Why are prospectors described as travelling down the Yukon River, when they are travelling northwards?

13. a) Why did the Mounties make prospectors take in a year's worth of provisions during the gold rush?

b) Imagine that you are a prospector back in 1896. List ten important items that you would include in your year's provisions.

c) As a prospector, you know there will be days when you will not be able to work, and when you may not have any money or gold to pay for entertainment. What items would you include in your pack to help you through those inactive times? Briefly explain your answer.

14. a) In what ways does tourism in the Yukon benefit from the activities of hikers such as those described in this account?

b) What costs are there for the region? Consider costs other than economic ones.

c) What types of tourist activities would likely yield the greatest returns with the least costs? Explain your answer.

Tourism in the Yukon

The annual value of tourism to the Yukon is about Cdn$37 million. Tourism creates employment in many service industries, particularly the hospitality sector and transportation. It is especially important to the Yukon because it is not as directly affected by the boom and bust cycles that influence resource-based industries. It is, therefore, vitally important that efforts are made to encourage tourists and to provide activities and services on which travellers are willing to spend money.

There are three major attractions for visitors in the region. The first consists of reconstructed places and activities associated with the gold rush. The second is the native heritage and traditions, and the third is the beauty of the largely unspoiled landscape, with opportunities to observe wildlife and to hunt and fish. As with any development, care has to be taken to ensure that tourist developments do not destroy the very beauty that attracts visitors to the area. Development should also not be to the detriment of residents who depend on the natural environment for their livelihoods.

Dawson City was the centre of gold rush activity, and the romantic, bawdy, and somewhat risqué image is being restored from amongst the ruins of the old settlement. People come to hear about and to experience life in those days when Klondike Kate, the toast of Klondike, and Charlie Meadows, owner and operator of the Palace Grand Theatre, were in their prime. They hear about the principal figures in the discovery of gold, whose names — Tagish Charley, Skookum Jim, and Kate Carmack — conjure up images of the excitement of the times.

Some of the most spectacular scenery in Canada is found in the Yukon Territory. This region has 21 mountain peaks that exceed 3300 m in height, including Mount Logan at 5951 m. Steep slopes, snow-capped summits, swirling rivers, and solitude await those hardy enough to pursue simple pleasures. Hiking trails have been established to encourage tourism, and the wildlife draws people who want to fish and hunt in the region.

Figure 13.10 *Robert Service's Cabin*
Service's cabin has been restored and is the site of poetry recitals, an example of a tourist development using local history.

Figure 13.11 *A Scene in Kluane National Park*
What problems might park officials face in dealing with tourism in this wilderness area?

The Yukon mountains are home to an unusual mixture of plants that have originated from the Arctic, western coast and mountains, the prairies of North America, and the steppes of Asia. Animals such as the Dall sheep, mountain goats, caribou, moose, grizzly and black bears are protected in Kluane Park, which is a UNESCO World Heritage Site.

15. You and a group of two or three friends are planning an outdoor expedition to the Yukon for one or two weeks.

    a) Decide in which school vacation you wish to travel, the kinds of activities in which you would like to participate, and the places you wish to visit.

    b) Make a list of equipment, food, and other essentials that you will need.

    c) Explain, in general terms, where you will obtain this equipment and how you will transport it from your home to the starting point of your expedition.

    d) How will you prepare for and cope with medical emergencies, both of a minor and major nature?

    e) Imagine that you have been on your journey. Write an account relating the events of one day on your trip. Your story should fit your situation and be imaginative.

Challenges for the Future

The basis for economic development in the Yukon has always been the exploitation of natural resources, particularly minerals. Until the region develops a more broadly based economy, it will be susceptible to periods of downturn in the resource industries, such as the one that occurred in the mid-1980s. In 1985, for example, earnings from the mineral industry amounted to only $60 million. Fortunately, demand rose the following year and the value of production increased to $180 million. By 1987, gold production achieved a 70-year record. In recent years, tourism has played a role in diversifying the economy, although the seasonal nature of the industry has limited its development.

Exploration and drilling activity off the coast of the Beaufort Sea offer no real relief from the focus on natural resources.

One of the side effects of a resource-based economy is the transient nature of a large part of the non-native population. Workers are attracted by high wages, but stay for only a short time and then return to their homes. The labour force of the region includes a disproportionate number of single men. Some companies have tried to counter this trend by offering assistance to married workers who move their families into the region.

The remoteness of this region makes it unsuitable for large-scale manufacturing. Local raw materials and suppliers are few, the labour force is small, and markets are distant. In recent decades, the government has attempted to improve connections by building roads and airports. The Dempster Highway is a good example of the region's "roads to resources" program. Finished in 1979, the highway links Dawson City to Inuvik in the Mackenzie River Delta.

The native people of the region have demanded a role in the development of the Yukon. They never signed treaties with the federal government and continue to press their case for more control of their resources. Full-scale economic development must wait until questions of ownership of land and resources have fully been resolved.

16. Suggest reasons to explain why frontier regions are almost always based on the exploitation of natural resources.

17. What are the five most important problems that must be solved before economic development can occur on a large scale in the Yukon? Record your ideas in a chart in which you note the problems, identify the ways in which the problems limit development, and suggest solutions to the problems.

Chapter Review

- Early economic development of the Yukon expanded rapidly with the discovery of gold.

- The gold rush ended when the mineral deposits were depleted.

- Boom-bust cycles are typical of economies based on natural resources.

- Tourism is a way of expanding the economy in ways that are less susceptible to fluctuations in the global economy.

- Tourism in the Yukon is centred on retelling the story of the gold rush, the native heritage and traditions, and the natural beauty of the area.

- Mining and tourism will likely remain the two most important economic activities in the Yukon for the foreseeable future, in spite of efforts to build a more diverse economy.

 ## Vocabulary Review

boom
bust
frontier region
hypothermia

pictograph
site
situation

 ## Thinking About...

THE YUKON AS A FRONTIER REGION

1. **a)** Identify several factors that make the Yukon different from other parts of Canada.

   b) What are some characteristics of the Yukon that make it similar to other frontier regions in the world?

2. In your opinion, what are the two greatest assets of the Yukon? Give reasons for your choices.

3. Imagine that gold had not been discovered in the Yukon in 1896.

   a) In what ways would the economic development of the region be different than it is now?

   b) In what ways might the lives of native people be different than they are now?

4. Compare the economic benefits of mining and tourism to the Yukon using a comparison chart. Some characteristics you might consider are:

   - number of jobs
   - salaries earned
   - days of employment per year
   - stability of jobs
   - ownership of companies

5. Plan an advertising campaign to attract tourists to the Yukon. Include in your plan:

   a) a clear goal statement (including the names of the countries that you would target with your advertising campaign)

   b) a resources list to help you achieve your goal

   c) problems that you will have to overcome

   d) actions that you would take to accomplish your goal

   e) an evaluation system

6. **a)** How does Figure 13.12 illustrate the boom-bust cycle?

   b) Describe the effects of a drop in the value of production for the Yukon. Give specific examples.

   c) Suppose you were a miner in the Yukon in 1985. What decisions might you have made about your future? Explain your answer.

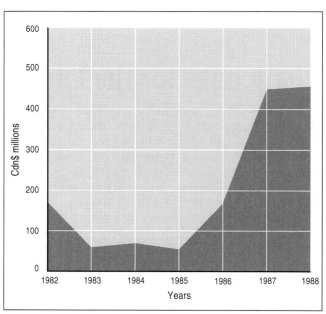

Figure 13.12 *Value of Mineral Production in the Yukon, 1982-1988*

Atlas Activities

1. Locate the Yukon on the World Physical map on pages A8-A9.

   a) Describe the landforms of this part of North America, giving names to the features where possible.

   b) Speculate on the forces that were responsible for creating the landforms of this area. Summarize your ideas in a short paragraph.

   c) In what ways would the physical make-up of this region influence its development?

2. **a)** Locate Dawson City on the maps on page A26. (See the more detailed map on page 206 for the exact location of this community.) Draw a climate graph for Dawson City by observing the patterns from the maps and climate graphs on this page of the atlas. Some points you should consider before you draw the climate graph are:

   - What is the average January temperature?
   - What is the average July temperature?
   - How much precipitation is this area likely to receive?
   - In what months will there be the most precipitation?

   b) Identify the three most important influences that shape the climate pattern of the Yukon. Give reasons for your choices.

3. Some people have suggested that the political boundaries of North America have been artificially created and do not reflect the geography of the continent. Using only the information on page A27, divide the continent of North America into three to seven countries that take into account the physical and economic patterns. Give your countries names and write a brief rationale for your choice of boundaries and of names.

 ## Further Explorations

1. Research the lives of some of the early prospectors. You might begin by reading Pierre Berton's *Klondike*. Write a short play, story, mime, poem, or song depicting some of the prospectors' experiences, or what you think they would have experienced.

2. Investigate the effects that global warming is likely to have on western Canada, particularly the Yukon.

   a) Summarize the findings of your research, using maps and graphs where possible.

   b) Speculate on the future of the Yukon, assuming a significant warming of the climate, a deteriorating agricultural base in previously fertile areas (such as the prairies), and a worldwide rise in sea levels.

3. Imagine that another significant deposit of gold was found on the banks of the Caribou River in northeastern Yukon.

   a) Locate the area on a map.

   b) Compare the accessibility of this gold deposit to that of the find near Dawson City. Consider both the physical location and the availability of transportation.

   c) Suggest the services and facilities that would probably be established. Draw a sketch map of the area showing where they might be located.

   d) What effects do you suppose would be felt in nearby settlements, for example, Inuvik and Dawson City?

   e) What legislation should the Yukon government enact to prevent destruction of the environment? Give reasons for your suggestions.

   f) In what ways should the interests of the native people in the area be protected?

   g) How should the Yukon government ensure that the whole territory benefits from the find? Explain your ideas carefully.

4. If you receive a newspaper regularly in your house, skim the business pages for several weeks to find articles that refer to the Yukon or to natural resource extraction in Canada. From these, select and explain phrases, sentences, or passages that illustrate the characteristics of frontier regions and their economies.

14

Researching Regions

Figure 14.a   *A Market in Tanarive, Madagascar*

Figure 14.b   *Fedala, Morocco*

Figure 14.c   *Peggy's Cove, Nova Scotia, Canada*

Figure 14.d *The Eiffel Tower, Paris, France*

Figure 14.e *Near Ayers Rock, Northern Territory, Australia*

GETTING STARTED

1. Brainstorm a list of at least 10 types of information that the photographs provide. For example, the ways people are dressed might provide information that would enable you to draw conclusions about climate or culture.
2. If you wanted to find out more about these places, what reference sources might you use?
3. How might you determine the potential for tourism in each of these places?
4. Which of these places appeals to you most as a tourist destination? Why?

LEARNING DESTINATIONS

By the end of this chapter, you should be able to:

- recognize the usefulness of regional study techniques;
- identify useful sources of information about places;
- draw tentative conclusions based on the information you have collected about places;
- communicate your conclusions in an appropriate manner.

Introduction

The regions in this book have been selected to enable you to learn about and appreciate the great variety of cultures, economies, and environments in the world. It would have been impossible to have included a detailed study of all parts of the world. However, the knowledge and skills that you have gained from the case studies in this book will enable you to better understand any region that you might study or visit in the future. The case studies have shown that the nature of any region depends upon a combination of many factors, including its:

- physical setting
- resources
- climate
- historical development
- culture and traditions
- political and economic systems
- outside influences
- economic development

By studying a variety of different regions and, even better, by visiting them with an inquiring and understanding outlook, we form a more complete image of the world. This knowledge allows us to understand other people's values and attitudes. Worldwide understanding such as this can lead us towards better global relationships and, perhaps, towards worldwide peace.

Understanding regions, however, takes detailed planning and organization, careful data collection and analysis, and the drawing of thoughtful conclusions. In short, understanding regions requires good research skills. This chapter provides a model or process for research.

1. Look over the regions included in this book.

   a) Which region, in your opinion, contrasts the most with your own area? Give at least three reasons for your choice.

   b) If people from the region you selected in a) were to visit your own area, what might they find strange, or even frightening?

   c) What be some causes for people's feelings of concern or fright when they visit unfamiliar places?

2. If you were to write a regional study about your own area, what focus would you choose? Why?

Conducting a Regional Study

There is an old saying that goes: "You will not get to where you want to be unless you know where you are going." This is true in life, as well as in research. Having a clear direction and goal is the key to good research.

The first step in conducting a regional study is to determine which region to investigate. You may be assigned the region by the teacher, or you may choose a place yourself. The second step is to decide on an appropriate focus for the study. The focus gives a particular direction or range of topics to a case study. The focus may be suggested by the teacher, or you may be asked to determine for yourself what a suitable focus for the region would be. If the latter is the case, skim through a general reference book, such as an encyclopedia, in order to come up with a focus. Use this focus to make up an appropriate research question. Some possible questions for a regional study are: Does this region have tourist potential? How could living standards be improved in this region? Where might an iron smelter be located? In what ways is region A similar to or different from region B? Focus on just one research question — it will be more than enough!

Once you have established what your research question is, the next step is to determine types of information you will need to gather in order to answer the question.

Brainstorm topics that will fit the focus you have established. Suppose, for instance, that you choose to develop an economic question, such as: What types of economic development should this region pursue? In brainstorming, you might identify the need for information about educational standards, transportation facilities, average incomes, and so on. Organize your topics into categories, throwing out those that do not seem to fit or are inappropriate for the task. Five or six categories of information about the region may be enough to enable you to reach conclusions. Make up a form or method for recording information about each category. This **research organizer** will help show you what information you already have and what remains to be found. An example of an organizer is the chart on page 220.

Spend a few minutes planning your use of time. If working with others, divide up the work and establish deadlines for its completion. Use a **research log** to keep track of responsibilities and the work completed by yourself and group members.

RESEARCH CHECKLIST

Have you:

- considered the purpose of the research?

- identified a single research question, issue, or theme?

- decided on what types of information will be relevant?

- developed a method for recording the information?

- divided the research tasks among group members?

- set deadlines for the completion of tasks?

- started a research log?

---

3. In your own words, explain why it is important to plan and organize for research.

4. **a)** Turn back to one of the regional case studies in this book. Identify its focus. List some of the types of information provided in the chapter and show how they are grouped together.

   b) Suggest other information that might have been included in the chapter.

Finding Information About Regions

The method you selected for recording information about your region will guide you in particular directions. You do not want to record all information you find about a region, just information on the topics you have decided are important in answering your research question. Resist the temptation to record information simply because you have it; if it is not relevant, it is of little use to you. On the other hand, you might find important information you had not expected, and you may have to adjust your research organizer.

Information will come from a variety of sources: the classroom, the school's resource centre, and from outside the school. Ask for help if you are having difficulty in locating useful data. Don't overlook computers, video programs, and microfiche collections as sources of information.

As you are recording your data, judge its worth. Is it relevant? Does it tell you what you need to know? Is the data biased or prejudiced in any way? Have you got the complete picture, or just bits of evidence? Do all sources agree, or are there contradictions on key facts? If you are not satisfied that the information you have collected meets your needs, then look for new sources of information.

Since finding and recording information is time consuming, it helps to divide the work up among group members. Plan to come together frequently to check on the progress of each member. Compile the information you have collected to identify areas of greatest need and topics that are well documented. Keep your research log up-to-date.

RESEARCH CHECKLIST

Have you:

- used a variety of sources of information?

- used your research organizer to record information?

- frequently checked to see what information was still needed?

- assessed the information that you recorded for bias, accuracy, and completeness?

- continued to keep a research log?

CHECKBACK AND APPLY

5. List resources that are found in your classroom, in the resource centre, and in the community that could be used for regional studies.

6. a) In what ways might tourist information about places be biased?

   b) What are some clues you might look for to detect bias in information sources?

   c) How do you deal with the problem of biased information when doing research?

Using Your Information About Regions

The facts that you gather through research might be thought of as pieces of a jigsaw puzzle. Individually they make little sense; assembled in an appropriate way, the picture becomes clear. Facts are assembled in your mind. The picture becomes clear when you see broad patterns, connections between facts, and relationships among ideas. In Chapter 1, it was pointed out that a region is the sum of all the locations, patterns, connections, and relationships in an area (page 2). The region comes into focus once we have the right information and think carefully about it.

Once you have collected and analyzed all the relevant data, you will be in a position to draw a conclusion and answer your original research question. Making predictions on the basis of evidence is one way to apply what you have learned. Another approach is to make and defend a proposal for action, either in this region or in another situation. Making comparisons is also useful. The approach you choose will be determined by your original research question.

The results of research efforts should not be kept to yourself. Communicate what you have learned in some way, such as by making a display, writing a report, or giving a presentation. Be sure to share the conclusions you arrived at and the evidence that supports them.

Take a few minutes at the end of your study to evaluate the process (your research log will be useful for this). What went right with your investigation? Were you able to find the right types of information? What could be improved? Did everyone do their share of the work? What research skills did you improve? If you had to do the task over again, what would you change? Reflection like this will help you to improve on the next research task.

RESEARCH CHECKLIST:

Have you:

- examined the data closely for connections and relationships among the facts?

- recorded your conclusions?

- applied your conclusions by making predictions, drawing comparisons, or suggesting solutions to problems?

- communicated your findings to others in some appropriate way?

- evaluated your performance on this task?

- identified ways in which you might improve your research skills?

CHECKBACK AND APPLY

7. The analogy used to describe drawing conclusions was that of putting the pieces of a jigsaw puzzle together. What is another analogy that could be used?

8. Brainstorm ways in which the results of research could be communicated to:

- a friend
- classmates
- the general public

LOOKING BACK AT RESEARCHING REGIONS

Chapter Review

- Regions help us to understand the world by providing examples.
- Regional study techniques can be used to identify and answer important questions.
- Research must be carefully planned and organized.
- Data collection requires using a variety of sources and making judgements about the quality of the evidence.
- The results of research efforts should be applied in some way and communicated to others.

Vocabulary Review

research log
research organizer

Thinking About . . .

RESEARCHING REGIONS

1. a) State a purpose for doing each of the following parts of a regional study:
- planning the research
- evaluating sources for bias
- searching for connections and relationships
- assessing your performance

b) Produce a flow diagram to show the stages in the research process.

2. In what ways will researching a tourist region improve your enjoyment of a vacation in that region?

3. Design a research organizer that will allow you to gather information about tourist facilities in different regions. The data that would be collected using this organizer would be used to decide which destination would be most appropriate for a family of four, with two teenagers, vacationing for two weeks in the winter. Give reasons to justify your decision.

Atlas Activities

1. Collect information about four areas of the world to identify their unique characteristics.

a) Make a chart in your notebook with these headings:

Characteristics	Amazon Basin	Western Australia	Sri Lanka	Northwest Territories
climate				
transportation				

b) Use pages A4-A13 to identify characteristics of the four places. Record information for ten characteristics that clearly show the differences among the regions.

c) For each place, decide which focus would be most appropriate for a regional study, given the information in your chart. Give reasons to support your choices.

2. a) Use the physical map on pages A8-A9 to identify landform regions of the world. Use mountains, plateaus, and lowlands as categories. Mark the regions on an outline map of the world.

b) Describe the landform patterns that your map shows.

Further Explorations

1. Obtain several travel guides on various tourist destinations.

   a) List some of the features of the guides that help tourists enjoy places.

   b) According to the guides, what are some of the issues or problems in the places they describe?

   c) Do the guides give an unbiased view of the tourist area? Explain your answer.

   d) What other sources of information might you use to get a more balanced view of tourist destinations?

2. Examine several posters promoting tourist areas or facilities.

   a) For each poster, identify the message that it is intended to convey.

   b) Identify the types of people for whom each poster would have the greatest appeal.

   c) Explain the ways in which the posters create images in the minds of potential visitors.

   d) In what ways might posters present a biased or inaccurate image of a place?

Glossary

This glossary contains the key terms that appear in bold face type in the text, together with other terms useful to students of regional geography.

accessibility — This is a measure of the ease with which a place may be reached. Places that are easy to reach or get at are accessible. Good transportation routes improve accessibility.

air traffic controller — This is an individual who works at an airport and helps to coordinate the arrival and departure of all flights.

alluvial deposits — Sand and/or mud that is deposited by rivers are alluvial deposits. The deposits form deltas at the mouths of some rivers.

American plan — With American plan, the room rate includes breakfast, lunch, and dinner.

Amerindian — This term describes Native people who are descendents of the original inhabitants of North and South America.

annexation — Growing cities may take over by adding to their size from the surrounding rural areas.

aquifer — Underground rock that will allow water to pass along it is called an aquifer. Wells dug or drilled into an aquifer will contain water.

archipelago — This term refers to a group of islands. The islands in the most northerly part of Canada are called the Arctic Archipelago.

barrios — Barrios are slum communities that are home for Spanish-speaking peoples in North, Central, and South America.

bed and breakfast — This type of accommodation is often in a private home. It includes a room to sleep in as well as breakfast. It is usually less expensive than motels or hotels.

bilharzia — This is a disease of tropical countries usually associated with irrigated farmland. It is caused by a worm that burrows through a person's skin, usually the feet, and enters the bloodstream. It results in great weakness and an inability to do much work.

bituminous coal — This is a moderately hard coal used in steel making.

boom — When economic conditions are good and jobs are increasing in number, a place is having a boom time. The opposite of this is a bust.

bust — This is a time of poor economic conditions and high unemployment. Bust and boom times are often cyclical, particularly for places dependent on primary industries.

capital — Capital is wealth that can be used for investment in businesses.

carnivores — Flesh-eating animals, such as cats, raccoons, and seals, are carnivores.

charter flight — A flight booked exlusively for the use of a certain group of people is called a charter flight. It is usually cheaper than regular scheduled flights.

city-centred region — Regions that are dominated by large cities fit this category. Paris dominates the surrounding French countryside and is an example of this type of region.

component — In industry, this term refers to standardized parts that can be combined in different ways to produce goods that meet the specific needs of customers.

connection — The links or associations that places have with other places are called connections. Roads are obvious connections between places.

consensus — Consensus is a form of decision making that involves coming to general agreement so that all or most of the people accept the decision.

constitutional monarchy — In this form of government a hereditary ruler governs according to rules established in the constitution of the people; power is shared between the people and the monarch.

continental climate — This term refers to a climate that is modified very little by nearby water bodies. It receives small amounts of precipitation and experiences a wide range of temperatures.

conurbation — In some places, neighbouring communities expand until they join together, forming a conurbation.

country of first asylum — Faced with persecution in their home countries, many refugees flee to nearby countries where they can get temporary safety. In these countries of first asylum they apply to other places for permanent residence.

Creole — Creole is a language of the Caribbean that is based on two or more other languages, particularly European and African languages.

cultural region — A place that has a distinctive culture that sets it apart from other places may be said to be a cultural region. In Canada, Quebec may be said to be a cultural region distinct from English-speaking Canada.

culture shock — The difficulties that a person experiences when trying to adjust to the ways of life of a different country or area are referred to as culture shock.

decomposers — Decomposers break down plant and animal matter into the basic materials of which they are made, such as minerals and nutrients. These materials are used by living plants and animals in their natural processes.

delta — Deposits of sand and/or mud that collect at the mouths of some rivers are called deltas. Over time, these deposits may accumulate and form land that is only barely above the surface of the water.

demographic region — This term is used to identify regions with population characteristics that set them apart from other areas. Growth rates may be higher or lower or population densities may be different in these regions.

desalination — The process by which salt is removed from sea water is named desalination.

desert — There are both hot and cold deserts in the world. They are dry, barren areas with very little precipitation. In climatology, an annual average precipitation of less than 250 mm is considered to be a desert.

desert vegetation — Desert vegetation, such as cacti, succulent thorn bushes, quickly germinating seeds, is a sparse cover of plants that can survive in a very dry environment.

discretionary income — This term refers to the amount of income that remains after taxes have been paid and the essentials of life purchased. We make decisions about how this money is to be spent.

doubling time — Measured in years, doubling time is the period required for a population to double in size. The shorter the time, the faster the rate of population growth.

duty free — An item that can be purchased as one crosses an international border and that is exempt from government taxes is a duty free item.

eco-tourism — Some tourist activities are designed to allow visitors to experience natural settings with a minimum amount of damage to the ecosystem. This approach to planning and development is eco-tourism.

economic region — An economic region is a place that has distinctive economic characteristics, such as Florida (tourism) or the Prairies (grain farming).

entrepreneurs — This term refers to people who are willing to invest money in business ventures that may be risky.

Eurailpass — This is a special pass sold overseas that allows the holder to have unlimited first-class travel in many western European countries for a certain period of time. There are also passes available for students and children.

European plan — A room rate that is on European plan includes only lodging, no food.

Everglades — A large part of Florida is made up of a large swamp or marsh that is partly covered by tall grasses. This is called "the Everglades".

food chain — Energy moves through an ecosystem by way of the food chain. Producers use the sun's energy in plant growth; herbivores eat the plants; carnivores eat herbivores; decomposers break down all dead matter.

foreign debt — Foreign debt is money owed to the government or lending institutions of another country. Many poorer countries have borrowed heavily to finance development and now have staggering foreign debts.

foreign investment — This term refers to the money spent by companies or governments in business ventures in other countries. A Canadian business that built a factory in an African country would be engaging in foreign investment.

freeways — Freeways are highways that are designed to move large volumes of traffic long distances at high speed.

frontier region — In this book, Siberia and the Yukon are identified as frontier regions. The full potential of both places is not yet known, and physical constraints limit development. Settlement patterns and transportation routes are still poorly developed.

functional region — Regions that are centred on a particular place or activity are considered to be functional regions. Pizzas are delivered in a region that is centred on the pizza shop.

game reserves — Many countries have set aside areas where wildlife is to be left undisturbed. Hunting is prohibited in these game reserves.

gridlock — Gridlock most often occurs in larger, crowded cities. There are so many vehicles on the streets that little movement is possible.

gross domestic product — This is the total dollar value of all goods and services produced in a country in one year.

gross state product — The total dollar value of all goods and services produced in a state in one year is that state's gross state product.

herbivores — Herbivores are plant-eating animals such as cows, deer, and rabbits.

high season — This is the period of time when the greatest number of tourists travel to a certain area.

historic cultural region — A number of places in the world have developed cultures over thousands of years. This makes them distinct from places where the culture has been largely imported or in existence for a shorter period of time.

homogeneous — Groups where all the members share similar characteristics are homogeneous. A community in which most of the people have the same culture and ethnic background is a homogeneous community.

homogeneous region — This term is used to describe a region in which distinctive characteristics are common to all parts of the region.

hypothermia — Prolonged exposure to cold air or water can bring about hypothermia, an abnormally low body temperature. This condition can lead to death.

inclusive tour — An inclusive tour is a travel plan that includes transportation, whether by land, air, or sea, and accommodation, such as hotels and meals.

indentured labourer — At times in the past, workers signed contracts agreeing to work for specific periods of time, often for meagre wages. In return for their labour they learned a skill, obtained transportation to another place, or received food and lodging.

industrial nodes — These are centres of industrial concentration.

infrastructure — The infrastructure of a place includes such things as roads, electrical services, and sewers. These essential elements must be provided before a place can function well.

interdependent — Interdependence means that parts of an ecosystem are dependent on one another. For example, as deer become more plentiful, the number of wolves increases.

Interstate Highway — This term refers to high-speed American roadways that are intended to move people long distances quickly.

jet lag — This is a condition of fatigue and sleepiness that occurs after a long flight in an aircraft, particularly when a number of time zones have been crossed.

land reclamation — Land that was once under water is said to have been reclaimed.

leftist guerillas — Guerillas are fighters who carry on warfare by using sudden attacks and ambushes. Leftist guerillas hold socialist or Communist political ideologies.

lignite — This is a soft, brown coal that emits much smoke when burning. It is often used in the production of thermal power.

load factor — Load factor is the percentage of all the seats that an airline offers that are sold to paying passengers. It is an indication of the efficiency of the operation of that airline.

location — Everything occupies a particular place on the surface of the earth. By stating the locations, you are establishing the exact positions of the places.

maize — This is another word for corn.

mangrove tree — Mangroves are tropical, evergreen trees that grow in salt marshes and along coasts.

Mediterranean vegetation — Drought resistant vegetation, such as thorn trees, that is adapted to dry, hot summers and cool winters is called Mediterranean vegetation.

mestizos — People of mixed Amerindian and Spanish blood are called *mestizos*.

monsoon — This is the name given to seasonal winds centred on Asia. These winds reverse their direction with the seasons; they bring wet and dry conditions depending on their directions.

Montezuma's revenge — This condition of ill health is characterized by diarrhea and loss of body fluids brought about by the consumption of impure water or food.

mountain vegetation — Mountain vegetation is a sequence of vegetative types resulting from climatic differences at various altitudes.

multiplier effect — Money spent in an economy will have additional effects. For example, a shopkeeper will pay wages to clerks, who in turn spend some of it in grocery stores, and so on.

passport — This document, issued by a national government, certifies that an individual is a citizen of that country. Many countries require tourists to carry passports.

pattern — In geography, this refers to the arrangement of things on the earth's surface. Words used to describe patterns include concentrated, dispersed, linear, random, and so on.

permafrost — Permanently frozen ground is called permafrost. Only the top few centimetres will thaw during the summer months.

photosynthesis — This is the process by which plants make sugar from carbon dioxide and water in the presence of chlorophyl and light.

pictograph — This term indicates a form of communication using pictures as symbols to show information.

planning — This term is best applied to a method or process for achieving a desired goal. It involves thinking out beforehand how something is to be accomplished.

poaching — Poachers illegally kill or trap wildlife for food or profit.

polder — Polders are low areas that have been reclaimed from some body of water. They are usually protected by dikes.

political region — Areas that have distinctive political characteristics are designated political regions. In these places, the people see themselves as separate from others.

polyps — Polyps are simple water animals that grow in colonies, with their bases connected. Over time, their remains form coral reefs.

primate city — In some countries, individual cities become extraordinarily important, economically and culturally. These primate cities are many times larger than the other cities in the country.

producers — This is the name given to plants which convert the sun's energy into food which animals can then consume.

projection — A projection is a representation, upon a flat surface, of all or part of the surface of the earth.

pull factor — Conditions that attract people to a place are labelled pull factors. The security that people have in Canada would be considered a pull factor.

push factor — Conditions that encourage people to leave a place are called push factors. War and famine are important push factors.

region — A region is an area that has characterisitics that set it apart from other places.

relationship — The connections or dealings that one area has with another are called relationships. For example, the downtown business district of a town might serve as the commercial centre for the whole community.

research log — This is a method of recording activities during an investigation.

research organizer — The most efficient way of recording information is to determine beforehand what type of detail is needed. The organizer helps to structure the research activities.

ripple effect — The benefits of some economic activities spread out through a region like ripples on a pond. For example, when a tourist buys a locally made sourvenir, a number of people will benefit, including the shopkeeper, the artist, the government, and so on.

river-dependent region — This term refers to places that rely on the nearby river for economic activity, and indeed, life itself.

salinity — Salt water is described as being saline.

savanna ecosystem — Savanna is a term used to name grassland ecosystems in tropical or sub-tropical areas.

shifting cultivation — This is the practice of clearing land, using it for one or more years, then abandoning the area because it becomes infertile. The farmer then clears more land.

shoulder season — The off-peak times of the tourist season, next to the high seasons, are described as the shoulder seasons.

site — The site is the ground on which a place is constructed. In describing the site of a city, you would include details about soils, drainage, vegetation, and so on.

situation — The situation is the circumstances that surround a location. In describing the situation of a city, you would include details about transportation routes, locations of natural resources, nearby communities, and the like.

smog — Smog, which develops in large cities, is a combination in the air of chemical fumes and fog.

steppe — This term refers to a grassland ecosystem in southeastern Europe and Asia.

subsistence farming — Subsistence farming occurs when farmers produce just enough food to feed themselves and their families.

taiga — The evergreen forest of the subarctic, as in Canada and Siberia, is called the taiga.

terraces — In mountainous regions, the slopes have been made into a series of level surfaces or steps in order to create farmland. These steps are terraces.

tertiary institutions — Any educational facility beyond secondary school, such as college and university, is a tertiary institution.

theme park — Theme parks are amusement centres developed along particular topics, such as a medieval village or life in the future.

Thomas Cook — Thomas Cook was the first person to establish organized tours for the public.

time zone — The world is divided into 24 areas, each with its own standard of time. The zones begin and end at the International Date Line.

toll-free number — Businesses have toll-free numbers so that customers calling long distance do not have to pay for the calls.

tourist profile — A tourist profile gives basic information about a country that a tourist would need to know, such as climate, food, customs.

tour operator — A company that specializes in the planning and operation of pre-planned vacations, which are usually sold to the public through travel agents, is called a tour operator.

trade surplus — This occurs when exports (goods sold to other countries) are greater than imports (goods brought into your country).

transship — Goods which are transfered from one transportation method to another are said to be transshipped. At ports, goods may be transshipped from ships to rail or road vehicles.

travel agent — A travel agent is a company that sells travel tickets or services to the general public.

travel hub — This term refers to regional centres from which local transportation facilities serve nearby communities.

tropical rainforest — Tropical rainforests are tall, broadleaf, evergreen forests that grow in hot, wet climatic areas around the equator.

tundra — The tundra is the vast, treeless plain found in regions of high latitude. Mosses, lichens, and wildflowers are common forms of vegetation.

turnaround time — The time from design of a product until it is ready for sale to customers is the turnaround time.

urbanization — The large-scale movement of people from rural places to cities, and from small cities to large cities, is called urbanization.

visa — A visa is an official document or endorsement on a passport that allows the passport holder to visit a particular country. Some countries will not allow foreign travellers to enter without a visa.

wadi — This is a ravine or gully in parts of Africa and the Arabian Peninsula through which water flows during the rainy season.

INDEX

An italicized page reference indicates that there is a photograph on that page.

Atlantic Centred View

YUKON

LOS ANGELES

FLORIDA

THE CARIBBEAN

PERU

NETH